CRYSTALS AND X-RAYS

WITHDRAWN
548.83 2975

THE WYKEHAM SCIENCE SERIES

General Editors:

PROFESSOR SIR NEVILL MOTT, F.R.S.
Cavendish Professor of Physics
University of Cambridge

G. R. NOAKES
Formerly Senior Physics Master
Uppingham School

To introduce the present state of science as a university subject to students approaching or starting their university careers is the aim of the Wykeham Science Series. Each book seeks to reinforce the link between school and university levels, and the main author, a university teacher distinguished in the field, is assisted by an experienced sixth-form schoolmaster.

PREFACE

In this book we have tried to keep strictly to the brief given to us by Sir Nevill Mott—to provide an explanation of X-ray diffraction in elementary physical terms. This is fortunately possible; as Sir Lawrence Bragg has said, if the great optical physicists such as Fresnel and Fraunhofer were to return to Earth, we could quite adequately explain the subject to them in terms of concepts that they knew in their time 150 years ago. They would be completely lost with much of the rest of modern physics. They could not, of course, appreciate the full nature of X-rays, but given that these rays were waves of very short wavelength, they would be able to understand what the X-ray crystallographers were doing.

We have kept closely to optical principles, starting with the microscope and the limitation caused by the wavelength of light. X-ray diffraction then appears as one device for overcoming this limitation and extending the microscope to the world of atoms, through the interaction of X-rays and crystals. These two subjects are treated in outline only; we have sternly resisted the temptation to deal with them in depth.

With X-ray diffraction, however, we have given a rather more complete treatment than might have been expected in an elementary book; for example, we have made great use of the reciprocal lattice—a concept that some people regard as rather forbidding but which clarifies the whole subject so well, and has such a simple physical explanation, that we felt that it would be a pity not to use it.

The derivation of crystal structures is often regarded as essentially a mathematical subject. We, however, have stressed the physical aspects and have shown how these were applied to the first simple structures. The turning point came with the introduction of Fourier methods; we have explained the physical basis of these methods and have shown how they have resulted in the flood of results which shows no signs yet of abating.

We have also described the applications of the results of crystal-structure determination to the rest of science, and the ways that—even as a purely empirical tool—X-ray crystallography can be of use in Industry. We regard these points as important; fundamental physics must certainly be studied for its own sake, but it is gratifying if it can also help to solve other people's problems.

We hope that this book will excite some of the younger minds to discover for themselves the fascination of X-ray crystallography. But we also hope that even more experienced workers will find some subjects dealt with in a rather unusual way that will enable them to appreciate better the mathematical formulae that they use and the results that their computers obtain for them.

CRYSTALS AND X-RAYS

H. S. Lipson, F.R.S.—University of Manchester

WYKEHAM PUBLICATIONS (LONDON) LTD
(A MEMBER OF THE TAYLOR & FRANCIS GROUP)
LONDON AND WINCHESTER
1970

First published 1970 by Wykeham Publications (London) Ltd.

© 1970 H. S. Lipson, F.R.S. All rights reserved. No part of this publication may be reproduced, stored in a retrieval system, or transmitted, in any form or by any means, electronic, mechanical, photocopying, recording or otherwise, without the prior permission of the copyright owner.

Cover illustration—Photograph, with magnification of 10 000, of graphite crystal formation taken with scanning electron microscope. (Courtesy of I. Minkoff.)

ISBN 0 85109 150 4

Printed in Great Britain by Taylor & Francis Ltd.
10–14 Macklin Street, London, WC2B 5NF

TEMPLE
RUGBY
READING ROOM
WITHDRAWN

Distribution:

UNITED KINGDOM, EUROPE, MIDDLE EAST AND AFRICA
Chapman & Hall Ltd. (a member of Associated Book Publishers Ltd.), 11 New Fetter Lane, London, E.C.4, and North Way, Andover, Hampshire.

UNITED STATES OF AMERICA, CANADA AND MEXICO
Springer-Verlag New York Inc., 175 Fifth Avenue, New York, New York 10010.

AUSTRALIA AND NEW GUINEA
Hicks Smith & Sons Pty. Ltd., 301 Kent Street, Sydney, N.S.W. 2000.

NEW ZEALAND AND FIJI
Hicks Smith & Sons Ltd., 238 Wakefield Street, Wellington.

ALL OTHER TERRITORIES
Taylor & Francis Ltd., 10–14 Macklin Street, London, WC2B 5NF.

ACKNOWLEDGMENTS

I WISH to acknowledge the kind hospitality of the Departments of Materials Engineering, Physics and Chemistry of the Technion at Haifa, Israel, where most of my part of the manuscript was written during my tenure of a visiting professorship in 1969. Mr. Lee and I are also grateful to the three ladies who typed the text—Miss Margaret Allen, Mrs. Edith Midgley and Miss Valerie Flinn. Also Mr. Frank Kirkman's help in reproducing diagrams, taking X-ray photographs and producing optical illustrations has been invaluable, and Mr. Raymond Parkinson's reading of the entire manuscript has, I hope, helped almost entirely to eliminate any overlapping and inconsistencies.

ORIGIN OF FIGURES

Crystals by C. W. Bunn, 1964. Academic Press
Fig. 2.2, Fig. 2; Fig. 6.11, Fig. 16.

The Interpretation of X-ray Diffraction Photographs by Henry, Lipson & Wooster, 1953. Macmillan
Fig. 2.9, Fig. 26.1; Fig. 2.11, Fig. 26.2; Fig. 4.10, Fig. 208.2; Fig. 5.9, Fig. 43; Fig. 5.12a, Fig. 168.1; Fig. 5.12b, Fig. 168.2; Fig. 12.3, Fig. 127a.

Nature, Vol. 224, p. 492, Nov. 1, 1969.
Fig. 8.12, Fig. 2.

Crystalline State, Vol. I, by W. L. Bragg, 1933. Bell
Fig. 2.7, Fig. 24; Fig. 5.3, Fig. 9; Fig. 6.1, Fig. 32; Fig. 6.4, Fig. 165.

Crystalline State, Vol. IV, by Bragg & Claringbull, 1965. Bell
Fig. 9.7b, Fig. 117; Fig. 9.7c, Fig. 119; Fig. 9.7d, Fig. 120.

The Determination of Crystal Structures by Lipson & Cochran. 1966. Bell
Fig. 6.13, Fig. 112; Fig. 8.1, Fig. 210; Fig. 8.2, Fig. 197; Fig. 8.5, Fig. 207; Fig. 8.6, Fig. 205; Fig. 8.7, Fig. 211; Fig. 8.8, Fig. 217; Fig. 8.11, Fig. 229.

Optical Transforms by Taylor & Lipson, 1964. Bell
Fig. 10.3b, Fig. 30; Fig. 10.6, Plate 2; Fig. 10.7, Plate 37; Fig. 10.8, Plate 35; Fig. 10.8, Plate 35; Fig. 10.9, Fig. 136; Fig. 10.10, Fig. 143; Fig. 10.14, Plate 42.

The Great Experiments in Physics by Lipson, 1968. Oliver & Boyd
Fig. 3.15, Plate IV.

X-ray and Neutron Diffraction by G. E. Bacon, 1966. Pergamon Press
Fig. 4.2, Fig. 2.

Chemical Crystallography by C. W. Bunn, 1961. Clarendon Press (Oxford University Press)
Fig. 5.4, Plate VII; Fig. 7.6, Fig. 211; Fig. 10.12, Fig. 211.

Neutron Diffraction by G. E. Bacon, 1962. Clarendon Press (Oxford University Press)
Fig. 11.3, Fig. 9; Fig. 11.6b, Fig. 97; Fig. 11.6a, Fig. 107.

Optical Physics by Lipson & Lipson, 1969. Cambridge University Press
Fig. 5.11, Fig. 7.47; Fig. 7.7, Fig. 9.7; Fig. 13.1, Fig. 9.33.

Fifty Years of X-ray Diffraction by Ewald, 1962. N.V.A. Oosthoek's Uitgevers Mij, Domstraat 11–13, Utrecht, The Netherlands
Fig. 2.3, Fig. 4–4(2); Fig. 2.4, Fig. 4–4(1); Fig. 2.5, Fig. 4–4(3).

Nature of the Chemical Bond by L. Pauling, 1945. Cornell University Press
Fig. 9.10, Fig. 73.

Proc. Roy. Soc. A., Vol. 190, p. 474, Plate 8, 1947, by Bragg & Nye. The Royal Society
Fig. 9.12, Fig. 2.

Elementary Science of Metals by J. W. Martin. Wykeham Publications
Fig. 11.5, Fig. 2.15.

X-ray Diffraction by Polycrystalline Materials by Peiser, Rooksby & Wilson, 1955. Institute of Physics
Fig. 12.4, Fig. 166c.

Journal of the Iron & Steel Institute, Vol. CXLIX, No. 1, p. 134P, 1944. Iron & Steel Institute
Fig. 12.5, Fig. 6.

CONTENTS

CHAPTER 1

the microscope

1.1 *History of the microscope*

SINCE the main theme of this book is the exploration of matter on an atomic scale, it is necessary to begin with the first efforts to see detail beyond the scope of ordinary vision. The *simple microscope*, consisting of a single converging lens, is probably very ancient, a convex lens made from a single rock crystal having been found in the ruins of the palace of Nimrod (*c.* 860 B.C.) by the archaeologist Layard. That simple magnifiers were used in ancient times is the only logical explanation of the perfection of the minute detail in old carvings and the accuracy of the cut of gem stones. The use of convex lenses to improve the vision of long-sighted people can be traced back at least six hundred years.

Such lenses were, however, quite weak by modern standards, and some impulse was needed to produce the next great step—the manufacture of lenses of very short focal length. This impulse was probably the construction of the first compound microscope (§ 1.2), which is usually ascribed to either Hans Zansz or to his son Zacharias at the end of the sixteenth century. A *compound microscope*, made by the spectacle-makers at Middleburg in Holland and presented to Prince Maurice, was in the possession of Cornelius Drebell, mathematician to King James I, in the year 1617.

The invention of a method of grinding very-short-focus lenses is usually attributed to Leeuwenhoek of Holland. The nearer an object is to the eye the larger will be the image on the retina at the back of the eye, but the closeness is limited because normally the human eye cannot clearly focus objects within about 250 mm from the eye. It is much more useful to consider, as the criterion of size, the angle subtended by the object at the eye rather than the linear dimensions of the object. If a convex lens is placed in front of the eye (fig. 1.1) then an object can be clearly focused by the eye at a much smaller distance than 250 mm. The object then appears larger, and the angle subtended by the object at the eye is also larger in the proportion of the magnification. The shorter the focal length of the lens the nearer can the object be to the lens and still be in clear focus.

Historically the next great advance was the invention of the Wollaston doublet in which two plano-convex lenses were fixed a distance apart equal to the difference between their focal lengths. The plane sides of both lenses faced the object and the lens nearer the object had one-third the focal length of the lens nearer to the eye. The reason for the introduction of the Wollaston doublet was the realization that distortion of the image geometry became increasingly more apparent as the focal length of the lens in use became smaller. Single lenses introduced changes in the shape of the image which therefore did not give a real representation of the object. Straight lines, for example, became curved. Again, with short-focus lenses the variation of the focal length with colour became important.

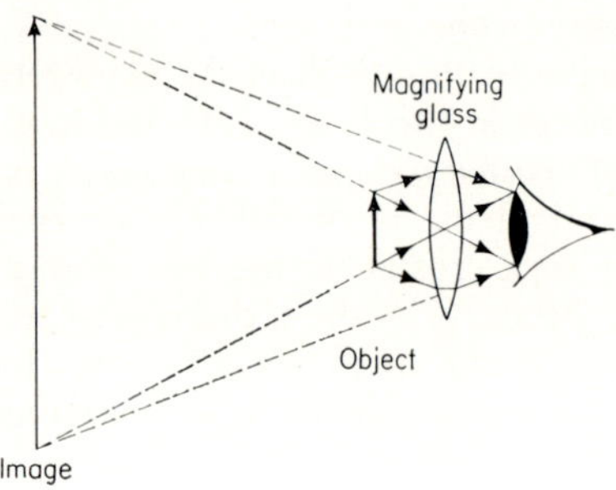

Fig. 1.1. Magnifying glass. The lens should be placed as close as possible to the eye and the object should be moved slowly towards the eye until a clear image is seen.

1.2 *The compound microscope*

Further progress in the examination of smaller and smaller objects depended upon the invention of the compound microscope, which in its simplest form consists of two short-focus convex lenses separated by a distance large compared with their focal lengths. Figure 1.2 shows, by means of rays, how this device produces a magnified image. The

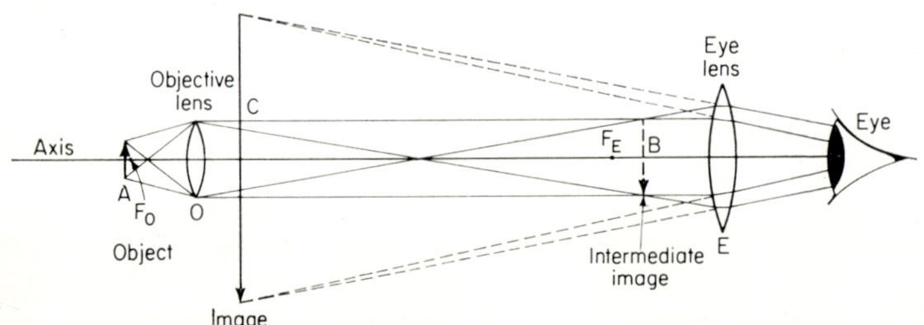

Fig. 1.2. Formation of image by compound microscope.

objective lens O gives a real, inverted magnified image at B of the object at A. This image is formed nearer to the eyepiece E than its principal focus F_E; in consequence a virtual magnified inverted image of the object is formed at C. F_O is the principal focus of the objective lens and F_E is the principal focus of the *eye lens*. In practice neither of these two lenses is a single piece of glass; they both consist of combinations of lenses designed and spaced apart to eliminate distortion in the image and to combine the differently coloured images of the object at the same place. In the usual use of the compound microscope the final image should be at the nearest point of distinct vision of the eye—that is, at a distance of about 250 mm from the eye.

1.3 *Improvements*

The final magnification produced by the compound microscope can be increased by three methods—shortening the focal length of the objective lens, shortening that of the eyepiece lens, and increasing the distance between these lenses. The older microscopes were made with draw tubes so that the distance between the lenses could be altered. Not only was this inconvenient, but the longer the microscope tube became, the smaller was the quantity of light passing through the objective and reaching the eyepiece.

During the eighteenth century Microscopical Societies became a gentlemanly vogue, and the members constructed their own microscopes. A genteel form of competition grew up, the comparison between microscopes being judged on what was called the Menelaus Scale, based on the pattern of lines on the scales of butterflies.

In a book published about 1860 the construction of an objective lens with a focal length of one-hundredth of an inch was reported. A microscope must have been needed to see this lens, because if it were a hemisphere of soda glass it would have had a diameter of one-hundredth of an inch. When these early microscopes are examined the minute size of the objective lens is at once apparent, the objective lens often being fitted into a lens holder with the central aperture about the size of a pinhole. Two distortions of the image, the one due to colour, and the other to defects of the lens, were being studied carefully at this time.

The defect due to colour, known as *chromatic aberration*, had long been known in telescopes. A combination of two lenses made of different kinds of glass, called an *achromatic lens*, was designed by John Dolland in 1757, but it was fifty years before such lenses were produced. The correction of the other defect, known as *spherical aberration*, was discovered by J. J. Lister, father of Lord Lister of antiseptic fame, in 1830. He found that every achromatic combination, with a plane surface towards the object and a convex surface towards the eyepiece, had two pairs of conjugate points for which the spherical aberration was corrected. The complexity of objective lenses can be realized when that designed by Lister and made by Ross was to consist of a triple front lens combined with two doublets; the focal length was one-eighth of an inch. The design of this lens was, however, so good that it was still being used a hundred years later.

1.4 *Depth of focus*

A further necessity in the progress of the microscope was simply mechanical; as magnification increased so the *depth of focus* decreased. Most of the readers of this book will have seen photographs in which the foreground and background are blurred because they are out of focus. If the normal eye has a least distance of distinct vision of 250 mm the table below gives the depth of focus for different magnifications.

Magnification	Depth of focus
10	2·5 mm
100	0·025 mm
1000	0·00025 mm

Thus, for very high magnification, very accurate and precise control of the fine adjustments of the microscope had to be designed—hence the coarse and fine adjustments on modern microscopes.

1.5 *Wave theory*

About the same time that the improvements in the optical and mechanical parts of the microscope were being developed, the theory of wave motion was being very much extended. An object in the path of a beam of light alters the beam so that information about the object is impressed upon the light waves; this is called the diffraction of the light waves by the object. A perfect image of the object can be obtained only if the whole of the information carried by the waves is used. This is impossible, of course, since lenses have finite sizes and so can accept only part of the complete wave system. Diffraction is directly caused by the wave nature of light, and consequently methods of dealing with the passage of such waves becomes necessary. The dominant feature of such methods is known as *Huygens' principle*, after the seventeenth-century scientist who stated it before the wave nature of light was fully accepted.

Huygens believed that light was a wave disturbance of some sort. If such waves proceed from a source, the surface that they reach at a given instant of time is called the *wave-front* (fig. 1.3). Huygens said that the wave-front at any later instant of time can be found by assuming that each point on the earlier wave-front acts as a point source of disturbance, from which spherical 'wavelets' spread out; the new wave-front can be found by drawing the envelope to these wavelets—the surface that is tangential to them as shown in fig. 1.3.

To apply Huygens' principle, it is necessary to know how to add waves. The simplest way is to use vector methods, each wave being regarded as a vector with its length proportional to the amplitude and its direction given by its *phase angle*. The phase angle is a measure of the displacement of the wave from a given origin: for example, we may take a cosine wave—disturbance = cos ωt, where ω is a constant—as having zero phase angle; the curve representing disturbance = cos $(\omega t-\alpha)$ then has a phase angle α, as we can see from fig. 1.4. Clearly, if disturbance = sin ωt, the phase angle is 90° or $\pi/2$ radians. Phase angles

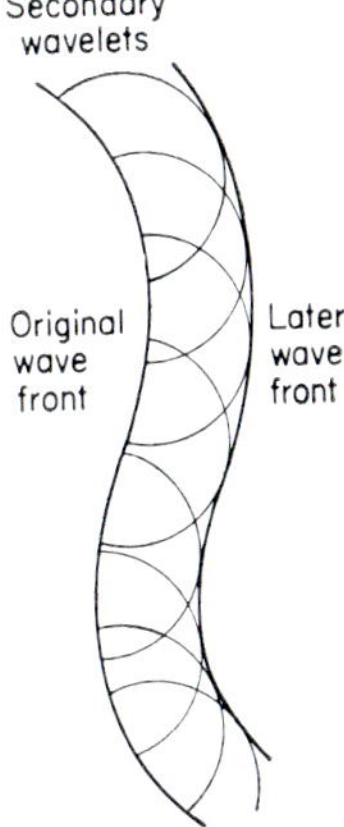

Fig. 1.3. An initial wave front and a later wave front produced from it.

are usually measured in radians, but they can also be expressed as fractions of a wavelength. Thus a sine wave can be regarded as being a quarter of a wavelength behind the cosine wave, and thus $\alpha = \pi/2$.

To add waves, we simply regard each as a vector, with length proportional to the amplitude and direction given by the phase angle. Figure 1.5 shows the resultant R obtained by adding three waves in this way.

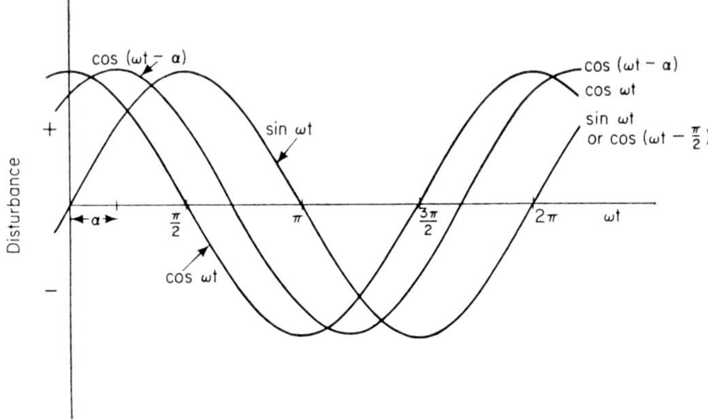

Fig. 1.4. Three sinusoidal waves, with an indication of the meaning of the phase angle, α.

1.6 *Diffraction by a slit*

To obtain the diffraction pattern of an aperture, such as a slit, we merely take a number of points regularly spaced within the slit—as many as we feel that we can handle—and regard each as a separate source. Let us start by taking only three points A, B and C (fig. 1.6).

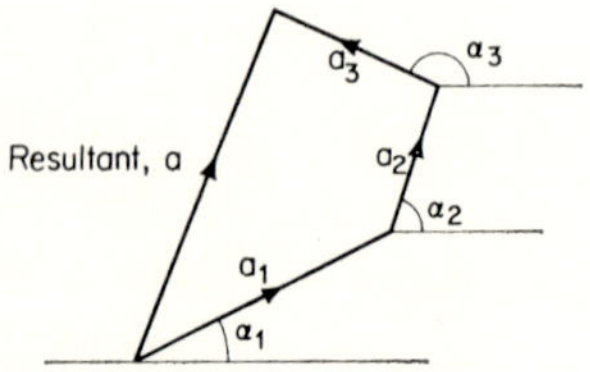

Fig. 1.5. Vector method of addition of three waves, with amplitudes a_1, a_2, a_3, and phase angles α_1, α_2, α_3.

Take A as defining zero phase angle. The three points produce equal disturbances, since we are assuming that the slit is illuminated by a uniform plane wave.

If the amplitude at some point E, at a large distance, is taken as a, then the disturbance produced at the point E can be found as follows. The wave along BE has travelled a distance BH further than the wave from A, so the angle between the vectors will be that angle corresponding

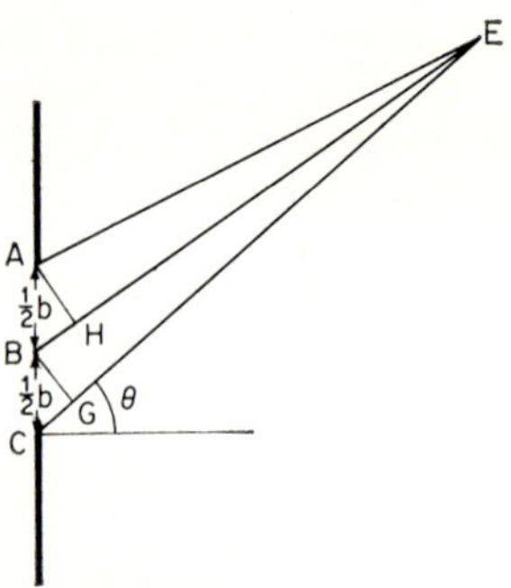

Fig. 1.6. Path differences of the waves proceeding from the points A, B, C, to the point E. E is considered to be far enough away from A, B, C, for the lines AE, BE, and CE to be effectively parallel.

to the distance BH. Now BH/$\frac{1}{2}b = \sin\theta$, and so BH $= \frac{1}{2}b \sin\theta$. Thus the phase angle separating the two waves is BH divided by the wavelength λ; this is equal to $(\frac{1}{2}b \sin\theta/\lambda)2\pi$ or $(\pi b \sin\theta)/\lambda$ radians. Similarly the amplitude of the wave CG will have a phase angle with respect to the wave form A of $\dfrac{2\pi\ CD}{\lambda} = \dfrac{2\pi b \sin\theta}{\lambda}$. Let us replace

$\frac{\pi b \sin \theta}{\lambda}$ by the symbol α. The vector diagram then consists of three lines of equal length, since A, B and C are of equal amplitude, each making an angle α with the previous vector (fig. 1.7). The resultant vector (i.e. the resultant disturbance) found by adding together the three waves AE, BE and CE is R.

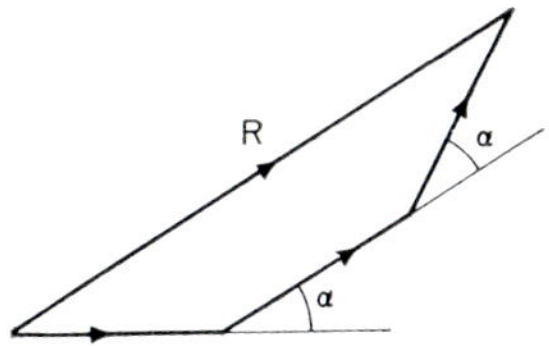

Fig. 1.7. The addition of three equal vectors with arithmetically increasing phase angles.

To progress towards Huygens' principle we now have to increase the number of sources of light in the slit AC from three to infinity. A little thought shows that the vector diagram now consists of the arc of a circle to represent the total amplitude of the wave motion coming from the infinite number of sources in the slit AC (fig. 1.8). The resultant vector

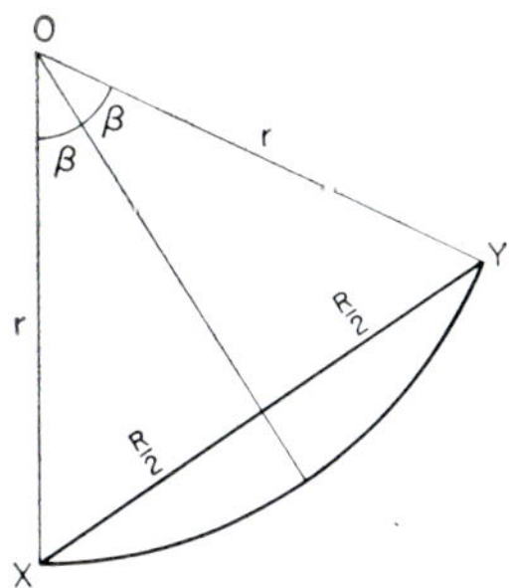

Fig. 1.8. The development of fig. 1.7 when a large number of small vectors is considered. The separate lines become the arc XY and the resultant R is the line XY.

in the direction θ is the chord of the circle joining the points corresponding to the sources at A and C. The arc XY represents the total light-wave amplitude entering the slit, and the chord XY (marked R) represents the resultant amplitude of the wave motion in the direction θ (fig. 1.8). Let O be the centre of the circle of which XY is an arc and let r be the radius of this circle. Then arc $XY/r = 2\beta$ radians, and $R/2r = \sin \beta$. Eliminating r between these two equations gives $R = \frac{(\text{arc } XY) \sin \beta}{\beta}$.

Thus the amplitude of the wave motion in the direction θ can be found. The intensity of the light is proportional to the square of the amplitude and so the intensity of the light in the direction θ will be given by $I = I_0 (\sin^2\beta)/\beta^2$, where I_0 = intensity diffracted at $\beta = 0$. If $\beta = \pm\pi$, $\sin \beta = 0$ and then I is zero. These are the first minima, and minima are repeated every time the angle β increases by π radians. This means that if a narrow slit is illuminated by parallel light, it produces a diffraction pattern consisting of a bright band in the centre with dark and bright bands alternately on each side. Since $\beta = \pi b \sin \theta/\lambda$ the first minima occur when $\beta = \pm\pi$, or $\sin \theta = \lambda/b$.

To produce an image of the slit the whole of the diffraction pattern produced by the slit must be used. But as we have said on p. 4, this is impossible; therefore the image is in some way different from the object. Since the single-slit diffraction pattern has its first dark band at an angle θ given by $\sin \theta = \lambda/b$, then the smaller b becomes, the larger will be the angle θ, and thus the gathering-together of the diffraction pattern will become more difficult.

1.7 *Abbe's theory*

The application, which we have just described, of the diffraction of light to the theory of the optical microscope was introduced by the German physicist, Ernst Abbe, about the year 1880. For the production of a truthful image of an illuminated structure by a lens it is necessary that the aperture of the lens be wide enough to transmit all the diffraction pattern that can be observed. High magnification, however, requires a lens with a short focal length, and because of this the lens must be small in size.

Porter, in 1906, devised an experiment to demonstrate the effect on the image of part of the diffraction pattern. A very brightly illuminated pinhole is placed in front of an achromatic lens which gives an image on a screen. If a fine wire gauze is placed in the path of the light the diffraction pattern produced on the screen consists of a number of patches of light—two main lines perpendicular to each other and some subsidiary lines at 45° to the main ones (fig. 1.9). If a hole is cut in the screen which allows only the central patch through, then the gauze cannot be seen through this hole. If a narrow slit is cut in the screen so as to transmit the horizontal patches, then it is possible to see through the slit the vertical wires of the gauze. If the slit is turned through a right-angle the horizontal wires can be seen, but the vertical wires have disappeared.

Porter also studied the effect on the image of an object caused when the light shining on it had to pass another object first. He fastened together two gratings of about 100 lines to the millimetre with the lines parallel and the gratings separated by 1 mm. Monochromatic light from a spectroscope illuminated the gratings, and the colour of the light used could be varied across the whole spectrum. The lines in the upper grating, on which the microscope was focused, were clearly visible on all colours except yellow, because, for this colour, the wires happened to fall in the positions of the dark bands. The diffraction pattern reaching the microscope carried information about both gratings, and the microscope cannot separate the two diffraction patterns from one another.

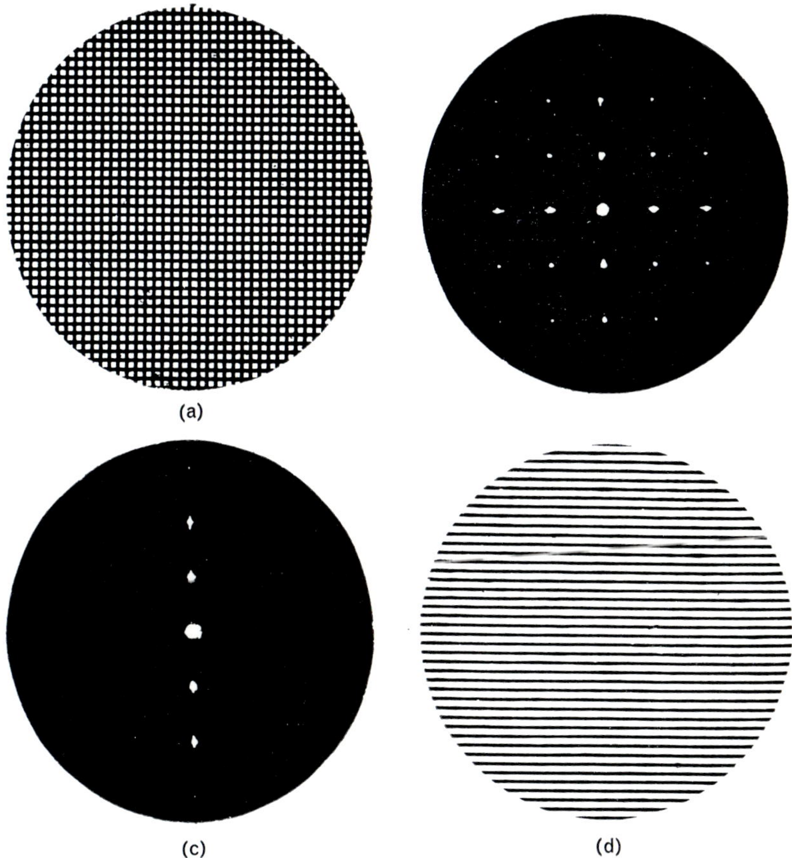

Fig. 1.9. (*a*) Image of gauze; (*b*) diffraction pattern of gauze; (*c*) (*b*) with all but centre row masked off; (*d*) image of gauze produced from (*a*), showing only the horizontal wires of the gauze.

1.8 *Resolving power*

We have so far discussed only the image of a single small object. If there are two or more small objects close together, then the diffraction pattern is profoundly affected by the presence of the neighbouring objects. The power possessed by an optical instrument of being able to give information about the presence of several objects close together is called its *resolving power*. The normal naked eye can resolve two points which are separated by about a tenth of a millimetre at the nearest point of distinct vision. It is easy to demonstrate this by making two slits in a card, putting the card in front of a bright light and then moving away from the card. At some distance it will be found impossible to decide whether there is only one slit or two. Since the

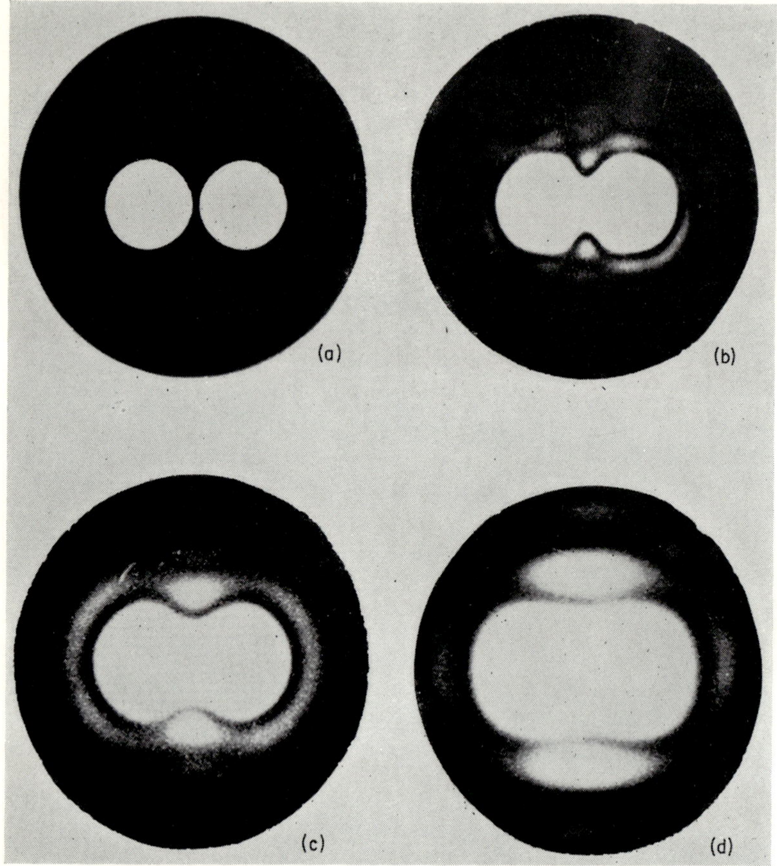

Fig. 1.10. (*a*) Image of two holes; (*b*) image at lower aperture, with resolution rather above Rayleigh limit; (*c*) still lower resolution, just about at Rayleigh limit; (*d*) complete lack of resolution.

images, as we have pointed out, are formed from the diffraction patterns, the two slits cannot be distinguished when their diffraction patterns are too close. A circular objective gives a diffraction pattern, from a point source, consisting of concentric circles alternately dark and bright around a bright central maximum called the *Airy disc*, after an Astronomer Royal who first worked out the pattern theoretically. The images of two circular objects are said to be separable if the centre of the principal maximum of one pattern falls on the first dark circle of the other pattern—the *Rayleigh criterion* (fig. 1.10). This is an extremely useful criterion in spite of the fact that astronomers and microscopists claim that they are able to distinguish between the image of a single object and that of two objects close together when the diffraction patterns overlap slightly more than the Rayleigh criterion.

Therefore it becomes obvious that it is useless to make efforts to increase the magnification if at the same time the resolving power of the instrument is such that one cannot tell whether one is looking at one object or several objects close together. As the diffraction pattern of a single slit would suggest, the larger the aperture of an objective lens the better the resolving power. Unfortunately, as we have shown, lenses of short focal length inevitably have small diameters. Abbe showed that the resolving power of the objective was proportional to the sine of half the angle i subtended by the object at the aperture of the lens. It is also proportional to the refractive index of the material between the object and the lens. This leads to a simple quantity for comparing objective lenses called *numerical aperture* (N.A.):

$$\text{N.A.} = \mu \sin i.$$

The resolving power d is then calculated by dividing half the wavelength of the light by the N.A., i.e.

$$d = \frac{\frac{1}{2}\text{ wavelength}}{\text{N.A.}}.$$

Amici (*c.* 1850) pointed out the increase gained in resolving power by placing a drop of water on the microscope slide and then bringing the

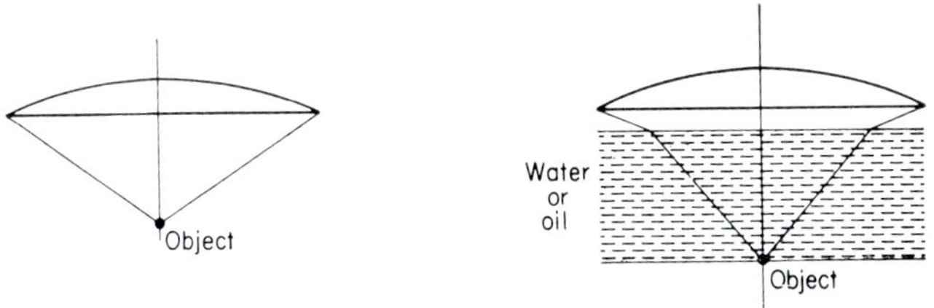

Fig. 1.11. Meaning of numerical aperture (N.A.).

objective lens down until the space between the object and the lens was filled with water. The refraction of light from water to glass makes the angle subtended by the aperture larger, and thus increases the N.A. (fig. 1.11). Since this is proportional to the refractive index of the liquid, higher resolution can be obtained by using liquids of high refractive index to match the glass of which the objective is made. This technique is called oil-immersion, and has a further advantage in that it results in an increased depth of focus.

1.9 *Methods of increasing resolution*

From the equation for the diffraction pattern of a single slit, giving the result that for the first minimum $\sin\theta = \lambda/b$, it is obvious that $\sin\theta$ becomes smaller either by increasing the aperture or by decreasing the wavelength. The preceding discussion has considered ways and means of increasing the aperture. Experiments to achieve greater magnification by using shorter wavelengths have also been carried out, particularly by Zeiss. A microscope using quartz lenses was made so that ultra-violet light could be used; the image had to be recorded by a camera since the eye is not sensitive to ultra-violet light. Greater resolving power and higher magnification were duly obtained, but the method is difficult and very expensive for the small increase in resolving power.

1.10 *Other wavelengths*

At this period the discovery of X-rays made a complete re-evaluation of the whole problem possible. It had become obvious that progress towards greater resolution and magnification must depend upon the use of the shorter wavelengths that X-rays provided. But, as we shall show in Chapter 10, the use of X-rays introduced other problems, and most of the rest of this book will be concerned with the way that these problems have been tackled and, to a large extent, overcome.

CHAPTER 2
X-rays

2.1 *Background to the discovery of X-rays*

The study of the discharge of electricity through a gas at low pressure was one of the main topics of interest in physics, and certainly the most exciting one, during the latter half of the nineteenth century. Geissler in 1855 had designed a vacuum pump that could produce lower pressures than any previously attainable and so made more detailed experiments possible. The study of the fluorescence of gases was initiated by Plücker in 1859 and was shown to be due to some sort of discharge or radiation coming from the cathode; it was soon found that the discharge could be deflected by a magnet held nearby but, whatever the deflection, one end of the discharge kept near to the cathode, thus indicating that it came from the cathode, not from the anode. Without a magnet, the discharge travelled in straight lines independently of the position of the anode; if this were placed in a side tube, the discharge passed by it.

These effects were all rather puzzling. Attempts to make measurements of electric fields and current densities in the discharge tube gave no help to solving the problem. It was not until 1897 that J. J. Thomson deduced that the fluorescence was caused by small negatively charged particles, now called electrons, travelling at extremely high speeds. But just before this date Röntgen made his great discovery of a new radiation produced by the cathode rays; since he did not understand the nature of this radiation, he called it X-rays.

2.2 *Röntgen*

Wilhelm Conrad Röntgen was born at Lennep in Germany in 1845, but his family left there when he was three years old to settle in Apeldoorn in Holland. He first attended the Van Doorn School in Apeldoorn and then the Technische School in Utrecht. In Utrecht he lodged with Jan Willem Gunning, Professor of Chemistry at the University of Utrecht under whose influence he went to Switzerland at the age of twenty to study at the Swiss Federal Technical School in Zürich. Three years later he graduated as an engineer. In 1868 he entered the University of Zürich to study for the doctorate of philosophy, and it was here that he met Professor Kundt, who had a profound effect on his future. In 1869 he presented a paper on 'Studies about Gases' as his thesis and was awarded his doctorate. Professor Kundt, who was lecturing in the University on the theory of light, offered the new Ph.D. a position as his laboratory assistant. Röntgen accepted, apparently having already decided that the intellectual atmosphere of a University was the answer to his own desires. The stay at the University of Zürich was short, as in 1870 Professor Kundt was invited to accept the Chair at the University of Würzburg in succession to Professor Kohlrausch and invited Dr. Röntgen to accompany him.

Röntgen was a meticulous practical physicist, and one of the first tasks that he undertook at Würzburg was to check the data published by Kohlrausch on the specific heats of gases. The results that he obtained differed from those of Kohlrausch, and in 1870 the *Annalen der Physik* published his corrections. Röntgen's instinct seemed to require him to check for himself the results of others, and it is clear that many papers on physics published at the time must have been studied very closely by him.

In 1872 Kundt left Würzburg for the Chair of Physics at Strassburg and again invited Röntgen to go with him; it was from here in 1875 that Röntgen was invited to be Professor of Physics at Hohenheim. The young professor remained in Hohenheim only one year, for he was then offered the Chair in Theoretical Physics at the University of Strassburg. After Strassburg Röntgen occupied the Chair at Giessen and then went to succeed his old friend Professor Kundt at Würzburg.

2.3 *Discovery of X-rays*

In 1894 Lenard succeeded in allowing cathode rays to pass into the air, through a thin metallic window, and his published papers aroused Röntgen's interest. Röntgen wrote to Lenard and obtained two of the thin metallic windows; with these he began the series of experiments which resulted in the discovery of X-rays.

In November 1895, Röntgen was trying to find out if cathode rays could penetrate the glass wall of the tube. He covered the tube with black paper to shut out stray light, and during the course of the experiment he noticed that a cardboard screen covered with barium platinocyanide crystals was fluorescing when the current was switched on. In a very short time he discovered that the effect was noticeable over a distance of several feet from the cathode-ray tube, and was therefore not due to the cathode rays penetrating the glass wall of the discharge tube; Lenard had found that the cathode rays could not traverse more than a few millimetres of air.

Alteration of the distance between the discharge tube and the barium platinocyanide screen proved that the brightness of the fluorescence was less intense at greater distances. Röntgen then tried the effect of placing objects between the discharge tube and the screen. A sheet of paper and a thick book produced little or no diminution in brightness. Aluminium sheets transmitted the effect, but a lead plate caused the fluorescence to disappear. Whilst holding a metallic plate between the discharge tube and the screen Röntgen noticed with amazement that the bones in his hand were visible in the shadow on the screen. A photographic plate wrapped in black paper was darkened by the radiation.

Röntgen now realized that he had discovered a new type of radiation, with remarkable properties. He called it X-rays. In 1913, for this discovery, he was the first person to be awarded the Nobel Prize for physics.

2.4 *Properties of X-rays*

Today, such a momentous discovery would have called for immediate publication. Röntgen, however, was much more cautious; he thought that if he delayed publication awhile he might be able to examine enough properties of the rays to establish their nature. He therefore spent about six weeks experimenting in his laboratory, hardly stopping for eating and sleeping, and working entirely alone. Only when he had satisfied himself that there was going to be no quick solution did he announce his discovery to the scientific world. It proved to be one of the most momentous announcements in scientific history. In these six weeks, Röngten made four important discoveries, all of which later led to new information about the rays. First, he examined absorption more systematically, and found that it was related to the atomic weights of the atoms in the absorbing material: platinum and lead absorbed more than silver and copper; aluminium absorbed hardly at all. The significance of this property becomes more apparent when we compare it with that of light; light is completely absorbed by thin sheets of light elements such as lithium and beryllium, but is transmitted by glass which may contain a heavy element such as lead. Clearly, X-rays are influenced by more fundamental properties of atoms than light is.

Secondly, X-rays affect photographic emulsions, and so X-ray 'photographs' could be taken. This property proved to be of great technological importance, but gave no evidence about the nature of X-rays.

Thirdly, they could cause electrified bodies to become discharged. This was a particularly subtle discovery, and one wonders how he came to think of the experiment. Scientifically the result was of great importance; we now know that it occurs because the air becomes conducting when X-rays pass through it—a phenomenon that we call ionization.

Finally, X-rays can be scattered by matter. That is, if a piece of material is placed in the path of X-rays, new X-rays appear in directions radiating from the piece of matter. Light also has this property, as we can see when a beam of light passes through dusty air; if light were not so scattered, we should not be able to see objects when they are illuminated.

But most of Röntgen's experiments were negative. X-rays were not reflected like light, nor were they refracted by prisms. None of the other properties of light, such as its diffraction, could be detected. The nature of X-rays remained a mystery, and when the discovery was announced it posed a problem that was not solved for another seventeen years. Were X-rays particles or waves?

2.5 *Nature of X-rays*

As is usual in science, when a subject has been opened by one man, others joined in and took it further than the originator. Walter and Pohl, in Germany, passed X-rays through a gold-plated tapering slit, 1/50 mm at its thin end, and obtained a suspicion of a blurring on a photographic plate; they deduced that, if the effect were diffraction, the wavelength of the radiation must be about one-thousandth of that of light. But the evidence was not sufficient to form the basis for any firm conclusions.

The ionization of air by X-rays was also investigated, particularly by W. H. Bragg, and this led to the opposite conclusion; the process could be understood only if the radiation were assumed to be particles. Scattering, however, seemed to support the wave theory; Barkla, in Liverpool, showed that the scattered intensity as a function of angle was similar to that of an unpolarized radiation, and that the radiation scattered through 90° appeared to be completely polarized. The results of his experiment are illustrated in fig. 2.1. He showed that the intensity scattered along the direction of the X-ray beam is greater than that scattered sideways, and that the sideways-scattered radiation cannot be scattered again normal to the plane of the diagram.

All this evidence was therefore inconclusive. Moreover, Barkla

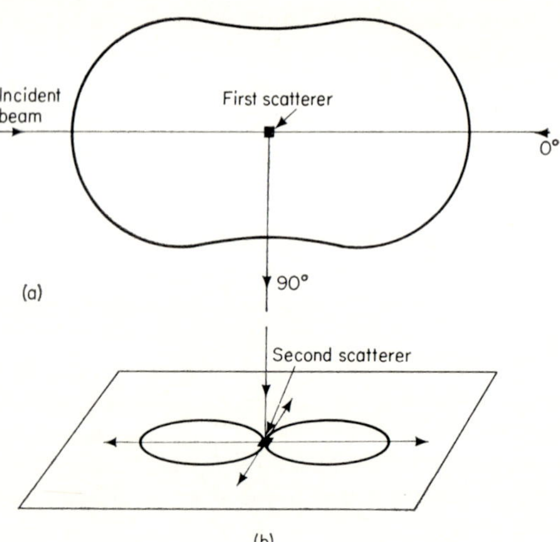

Fig. 2.1. Idealized polar diagrams of scattered intensity as a function of angle. (*a*) Directly scattered radiation; (*b*) secondary scattering of radiation first scattered through 90°.

made another important discovery; he showed, by absorption measurements using ionization as a measure of intensity, that if certain elements in the range chromium to zinc were used as anodes in X-ray tubes, the radiation had a component characteristic of the anode material. This result turned out to be of great importance later, and Barkla was awarded the Nobel Prize for it.

Thus the problem was left unresolved. There seemed no way of finding the answer, because the only definitive experiments—analogous to those in physical optics—seemed to be too difficult to carry out. If only diffraction gratings with spacings of a thousandth or even a hundredth of those of ordinary diffraction gratings could be made, the answer would be clear. But this seemed to be an impossible task.

2.6 *X-ray diffraction*

Nevertheless, the problem *was* solved in this way—but with a natural grating, not a man-made one. The complete story is a beautiful example of the way one scientist's mind can interact with another, with advantage to both.

In 1912 in Munich, a new research student, Ewald, was seeking the advice of a theoretical physicist, Laue, on the passage of radiation through a crystal. Laue knew nothing of the ideas about crystals, and was surprised to hear that they were considered to be formed by stacking units in a regular three-dimensional array (fig. 2.2). He asked about

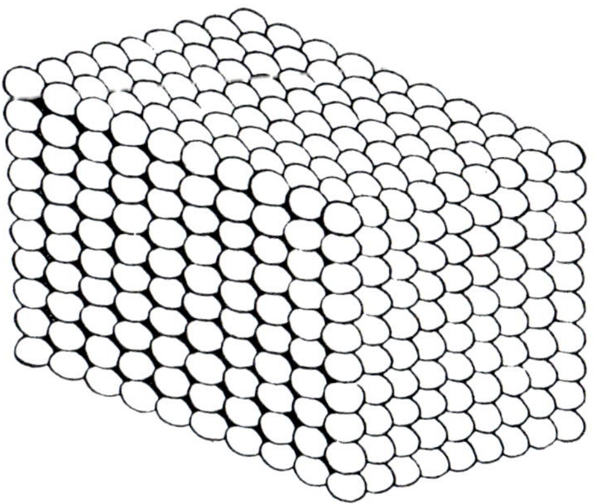

Fig. 2.2. Huygens' idea of the formation of a crystal of calcite by the stacking together of oval units.

the possible size of these units, and the answer that Ewald gave him suggested that crystals should be just right for diffracting X-rays. But he was a theoretical physicist and not used to apparatus, and so he enlisted the help of two experimental physicists, Friedrich and Knipping. They set up the apparatus shown in fig. 2.3, and after some failures they obtained the photograph shown in fig. 2.4—the first X-ray diffraction photograph. The diffracting crystal was copper sulphate. (It was only with some reluctance that they put the photographic plate in the path of the direct beam; it seemed to be the wrong place to look for diffraction!)

Fig. 2.3. Apparatus used by Friedrich and Knipping to explore the effects produced when a crystal is irradiated by a fine beam of X-rays. The X-ray beam travels horizontally from left to right, impinges on the crystal supported above the horizontal circle, and is diffracted on to the plate on the right-hand side.

One photograph does not, of course, prove anything. Friedrich and Knipping showed that different orientations produced different arrangements of spots on the photographic plate, and that different crystals produced different patterns. If the copper sulphate were powdered, the pattern disappeared altogether. Zinc blende, ZnS, gave particularly simple and beautifully symmetric patterns if the crystal were correctly oriented (fig. 2.5).

Thus the problem was solved. X-rays *were* waves. W. H. Bragg's ionization results gave some cause for worry, but even his most ardent supporters were ultimately convinced. (Now we know that he also was right; according to the quantum theory, a radiation can act as if it is composed of particles.) The next step was to try to understand the phenomenon in more detail to see whether it could be used to advance our knowledge of crystals. Here difficulties arose.

2.7 *X-ray diffraction theory*

Laue's theoretical ability stood him in good stead; three-dimensional gratings were more difficult to cope with than one-dimensional gratings,

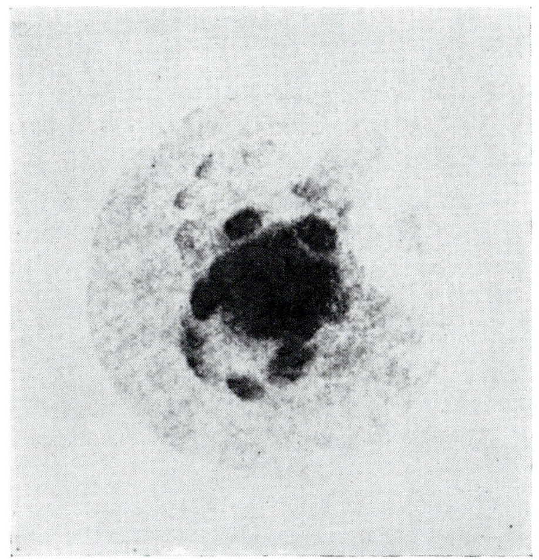

Fig. 2.4. The first X-ray diffraction photograph.

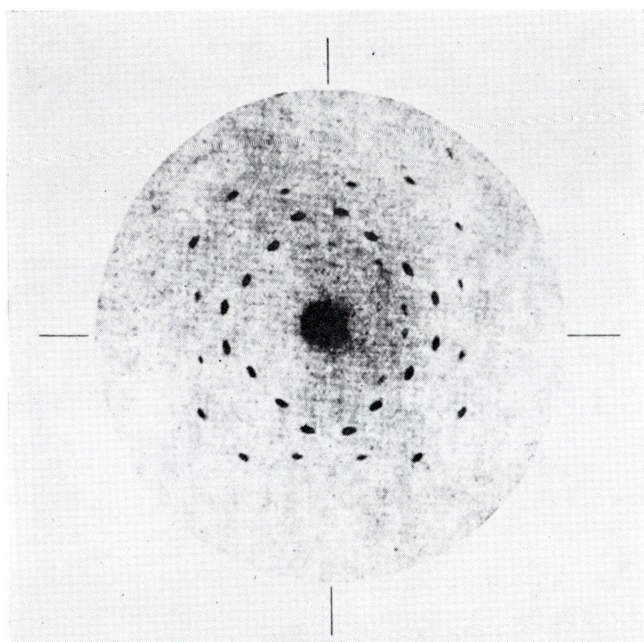

Fig. 2.5. An early X-ray photograph obtained by passing a fine beam of X- rays along an axis of symmetry of a crystal of zinc sulphide.

but the theory is essentially the same. But this theory did not work. Only a few of the spots on the ZnS photographs were explainable if one assumed that the X-rays had a single wave-length λ; five wavelengths had to be introduced to account for most of the diffraction patterns. This assumption seemed to Laue to be unnatural, and he had to confess defeat.

Progress in fact came from quite a different approach, introduced by W. L. Bragg—W. H. Bragg's son. The shapes of spots obtained with the photographic plate at different distances from the crystal suggested to him that the X-rays were somehow or other being *reflected* from plane mirrors and this idea worked. The mirrors were planes of atoms spaced equidistantly, and the equation giving the angles at which diffraction takes place is the well-known Bragg equation:

$$n\lambda = 2d \sin \theta.$$

Here n is an integer, d is the spacing of the planes and θ is the *grazing* angle of incidence of the rays on the lattice planes (fig. 2.6).

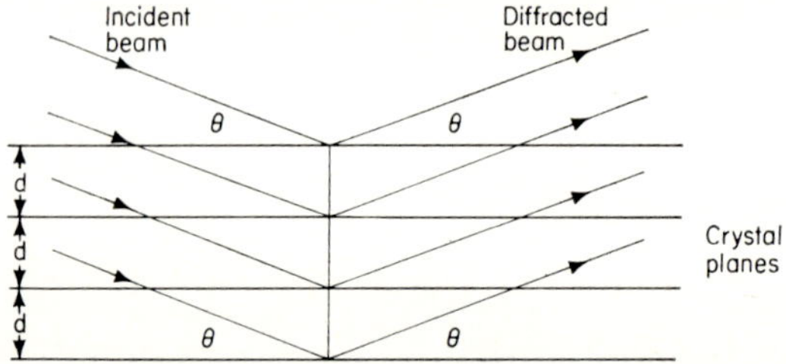

Fig. 2.6. Significance of the symbols in Bragg's law.

This equation will be explained more fully in Chapter 4. For the moment we wish merely to use it to obtain a deeper insight into the nature of X-rays. But there is a further problem to be solved: we know neither λ nor d. With a man-made grating we know d and can therefore find λ, but Nature makes crystal gratings and does not tell us what the spacings are. The solution was obtained by W. L. Bragg when he worked out the first crystal structure, and his method will be described in detail in Chapter 6.

Laue and the Braggs received the Nobel Prize for their work. X-rays were truly a rich fishing-ground for Nobel Prizes, but they went only to the people who knew how to fish properly!

2.8 *Further information about X-rays*

There are two questions that pose themselves. Why had Laue's theory been unsuccessful and what was the nature of Barkla's characteristic radiations? Bragg's approach gave the answers to both these questions. For, although Bragg's equation looks like the ordinary diffraction-grating equation, it is physically quite dissimilar. For a fixed wavelength, it is obeyed only if θ happens to be one of the solutions to the equation

$$\theta = \sin^{-1} n\frac{\lambda}{2d};$$

for any other angles, no reflection takes place. Let us take $n = 1$, for example. Then θ if fixed, and thus, if the rays are not incident at this angle on the reflecting planes, they will not be reflected. As the crystal is rotated, an orientation is reached at which θ is correct, a reflection will flash out, and it will disappear again when the angle is changed further. The same sequence of events will occur for $n = 2, 3, 4 \ldots$, etc.

Suppose, however, that the X-rays contain a range of wavelengths, like white light. Then, for a particular angle of incidence on the crystal planes, a specific wavelength satisfying Bragg's law will be reflected. As θ changes, λ also changes, and therefore varying the angular position of the crystal is equivalent to sweeping across the spectrum of the radiation. Based upon this idea, the Braggs built what they called an *ionization spectrometer* (fig. 2.7). With a crystal of rock

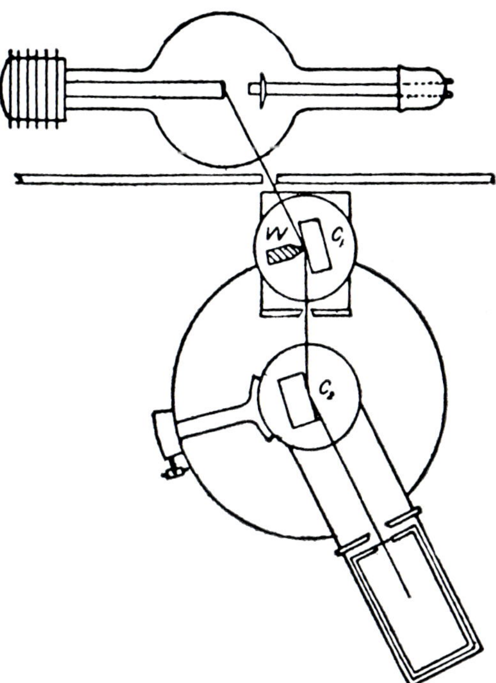

Fig. 2.7. Ionization spectrometer.

salt and an X-ray tube having an anode of palladium they found a spectrum resembling that shown in fig. 2.8; this showed a background with a continuous distribution, like that of white light, and superimposed upon it were two strong lines. The complete story was now clear.

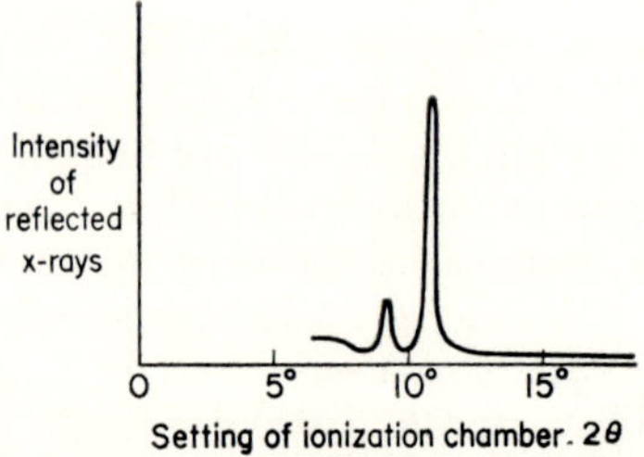

Fig. 2.8. An early X-ray spectrum from a target of palladium, showing a continuous band of radiation with α and β lines superimposed.

Laue's theory was incomplete because he had not thought of the possibility of a continuous distribution of wavelengths in the X-ray beam. It is odd that, having increased the number of wavelengths from one to five, he did not take the logical step of increasing it to infinity! This radiation is called *white radiation* because of the similarity to white light.

The strong lines are Barkla's *characteristic radiation* (§ 2.5) and the

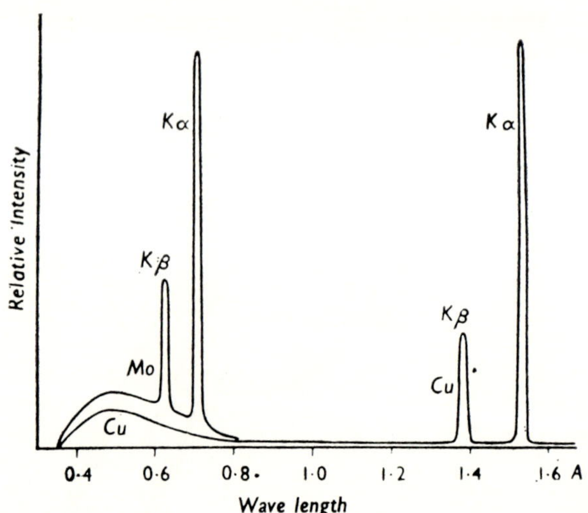

Fig. 2.9. Spectra from targets of copper and molybdenum, with a p.d. of 35 kV. The wavelengths are measured in Angstrom units (Å) where 1 Å $=10^{-10}$ m.

Braggs were able to measure the relative wavelengths of the lines from several elements. Moseley, in Manchester, went even further and constructed an X-ray tube in which different anodes were presented to the electron beam; he was able to show the succession of X-ray wavelengths from elements in order in the periodic table, and so established firmly the concept of atomic number. If he had not died in the First World War, he would almost certainly also have been awarded the Nobel Prize for his work.

More recent work on X-ray spectra gives results of the form shown in fig. 2.9. There is a short-wavelength limit set by the maximum energy that the exciting electrons have; the higher the tube potential, the shorter is this limit. The two lines are called β and α in order of increasing wavelength. Both are complicated lines, but to a first approximation β can be regarded as single and α, which is stronger, has two components, α_1 and α_2, whose intensities are in the ratio of 2:1.

Almost everything was now ready for use, except the scale of wavelength. As we have pointed out, we shall have to wait until Chapter 6 to see how this scale was established.

2.9 *X-ray tubes*

Modern X-ray tubes look quite different from the early ones that were made; just as the first motor-cars were adapted from the shapes of horse-drawn carriages, so the early X-ray tubes looked like discharge tubes. The tube with which Röntgen discovered X-rays was a rather bulbous affair with the anode in one side (fig. 2.10 *a*). The next step was the interposition of a definite *target* or *anticathode* to intercept the cathode rays, resulting in a tube like that shown in fig. 2.10 *b*. Also the surface of the cathode was curved; the cathode rays seemed to start in paths normal to the surface, and so a focusing effect could be produced: the X-rays emerged from a small area, called the *focus*, on the target.

An unexpected difficulty arose with these early tubes: the pressure inside did not remain constant, but it tended to *decrease*, not to increase as might have been expected. Apparently the residual gas adsorbed on the walls under the influence of the discharge. Various ingenious devices were made to try to let gas into the tube if it became too 'hard', as the effect was called; an example is shown in fig. 2.10 *c*. But none of these devices was really satisfactory, and X-ray tubes remained difficult things to control.

In 1913, however, the ideal solution was proposed. The American physicist, Coolidge, made a tube in which the pressure was as low as could be obtained, and electrons were produced from a heated tungsten filament (fig. 2.10 *d*). Despite a rearguard action from certain people who, with some justice, objected to the contamination of the target by tungsten from the filament, the *Coolidge tube* has now replaced the so-called *gas tube* completely.

X-ray diffraction, which required long exposures, brought in the need for compactness. The bulbous shapes of the early tubes meant that the recording apparatus had to be a long distance from the focus. So new shapes of tube arose with metal ends, glass being used only for insulation (fig. 2.10 *e*); these metal ends also facilitated water cooling and thus allowed much greater powers to be used. A modern form of tube is shown in fig. 2.10 *f*.

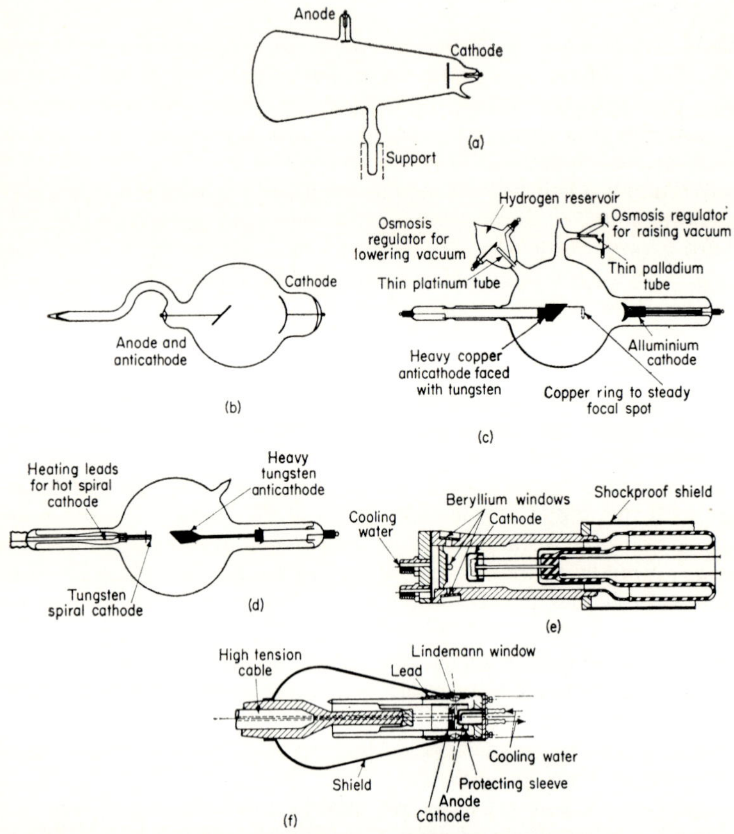

Fig. 2.10. The evolution of the X-ray tube. (*a*) Röntgen's tube; (*b*) tube with curved cathode; (*c*) tube with device for 'hardening' and 'softening' the vacuum; (*d*) Coolidge's hot-filament tube; (*e*) metal tube with glass insulator; (*f*) modern X-ray tube made by Philips, Eindhoven.

But with all these advances, we must remember the method of the production of X-rays has not changed; we still bombard a piece of matter with high-speed electrons. This is equivalent to playing a piano by dropping stones on it! It is an extremely wasteful process; only a small fraction of 1% of the energy is converted into X-rays, and only a small part of this forms the characteristic radiation that we particularly want. A spectacular advance would be made if we could find a way of exciting the electrons individually, as we play on the keys of a piano. But there seems to be no hope at all of carrying out this suggestion. Röntgen's original method is still the only one possible.

2.10 *Origin of X-rays*

Although for the purposes of this book we do not need to know anything about X-rays other than that they are an electromagnetic radiation whose wavelength we can measure, it is obviously unsatisfactory not to understand as fully as possible what is going on in an X-ray tube when the rays are being produced. We shall therefore now describe the main outlines of the processes involved.

The basic fact is that the rays are produced when electrons decelerate. In its crudest form, when a moving electron is stopped suddenly all its energy appears as a *quantum*, $h\nu$, of X-rays; ν is the frequency and h is Planck's constant. Now $\nu = c/\lambda$, where c is the velocity of X-rays—and of light—and the energy of an electron of charge e in falling through a potential difference V is eV; thus

$$\lambda = \frac{hc}{eV}. \tag{1}$$

In general an electron will not lose all its energy in this way; it will strike a number of glancing blows on the atoms that it strikes, and the main effect will be to cause them to vibrate and so to increase the temperature of the target. Equation (1) therefore gives the minimum value that λ can possibly have, accounting for the short-wavelength cut-off in fig. 2.9. Longer wavelengths are more probable and so the rapid rise in the curve of fig. 2.9 arises. There is no upper limit, accounting for the gradual fall-off in intensity at much longer wavelengths. In this simple way we can account for the main features of the spectrum of white radiation (§ 2.8).

But what of the characteristic spectrum? This is produced in quite a different way. Sometimes the impinging electrons can make a direct impact upon one of the inner electrons in an atom of the target and, if the energy is great enough, can knock it right out of the atom. The atom is then unstable, and another electron in the same atom will drop into the space vacated. In so doing it loses energy and a quantum of radiation is emitted.

If E is the energy lost we have a result similar to equation (1):

$$\lambda = \frac{hc}{E}. \tag{2}$$

E is a definite quantity associated with the particular energy change in the atom, and so the wavelengths concerned are specific; several wavelengths are possible and they constitute the characteristic spectrum (§ 2.8).

The process is similar to that of the production of light from a discharge tube, but for light the outer electrons only are involved; for X-rays, the tightly bound inner shells—K, L, M—are concerned (fig. 2.11). If a K electron is dislodged, and an electron falls from the L

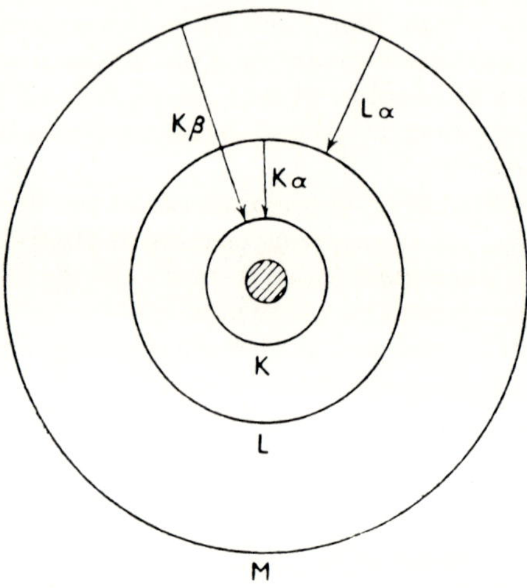

Fig. 2.11. Electron orbits in an atom, showing transitions associated with the production of characteristic radiation.

shell to the K shell, the Kα X-ray line (§ 2.8) is produced; if an M electron falls in—a less likely occurrence—the weaker Kβ line is produced. The letter K is used to indicate the shell into which the electron falls. These are the main spectral lines used in X-ray diffraction work. If an electron falls from the M shell to the L shell, a much longer wavelength is produced, but this is not much used in practice.

There are still some very weak lines in X-ray spectra that have not been properly accounted for, and there is no full theory of the shape of the spectral distribution of the white radiation, but on the whole the production of X-rays is probably one of the most clearly understood processes in modern physics.

CHAPTER 3
crystals

3.1 *History*

THE symmetry of shape, smoothness of surface and the colour or brightness of naturally occurring crystals have interested man from time immemorial. Fluorite crystals from the Blue John mine in Derbyshire were valued by the Romans as decorative objects. Diamonds, emeralds, rubies and sapphires have been of great value from very early times. The size of crystals found naturally as minerals varies between crystals of beryl discovered in America 1·20 metres long and 0·6 metres thick and weighing about 5 tonnes, to tiny, almost microscopic diamonds found in many parts of the world.

The same chemical compound occurs as differently shaped crystals; sometimes, for example, fluorite is found as a cubical crystal, but occasionally octahedral crystals are discovered. Diamonds, although mostly found in a pebble-like form due to the action of water, do occur as regular octahedra, whilst artificially produced diamonds are cubic in shape (fig. 3.1). Very many of the naturally occurring crystals can

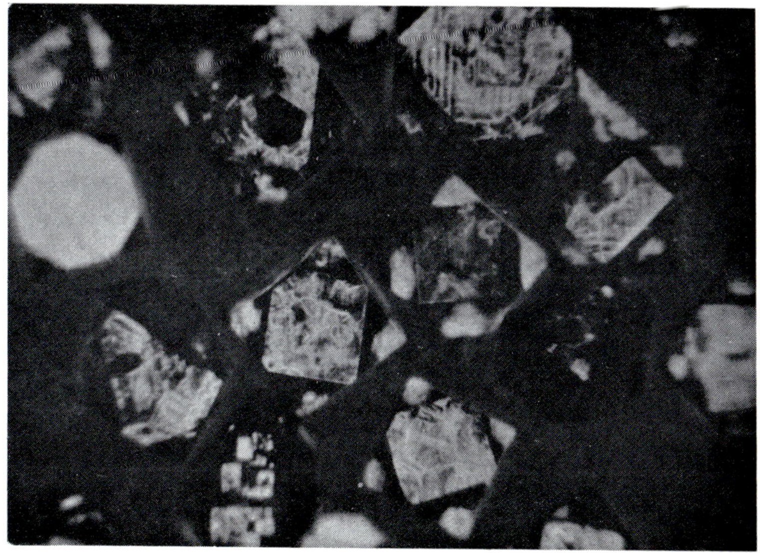

Fig. 3.1. Some artificial diamonds, which are cubical in shape. (Magnification 30 ×.) (Courtesy of M. Seal, International Diamond Centre.)

now be manufactured. The ruby-like jewels in watches are produced on a large scale from aluminium hydroxide, which is melted and allowed to fall on a small single crystal—what is called a 'seed' crystal—in an evacuated enclosure. The crystals of silicon used for transistors are made by taking a rod of the element and heating it near one end by means of a small furnace so that a small region just melts but holds together by surface tension; the furnace is them moved slowly along and the molten zone follows with it, usually leaving behind a good single crystal. This method has the additional advantage that impurity atoms tend to be carried along with the molten part, the resulting single crystal therefore being very pure indeed. The process is called *zone refining*.

Another important characteristic of crystals is *cleavage*. Most crystals, when hit with a hammer, break with irregular fragments, but some form small crystals with plane faces. If we carry out the more controlled experiment of placing a knife edge on the surface of a cubical crystal of rock salt, NaCl, with the edge parallel to a cube face, a sharp blow on the back of the knife blade will cause a flake to break off; the flake has its exposed face—the *cleavage face*—exactly parallel to the original face (fig. 3.2). In this way rock salt can be converted into a

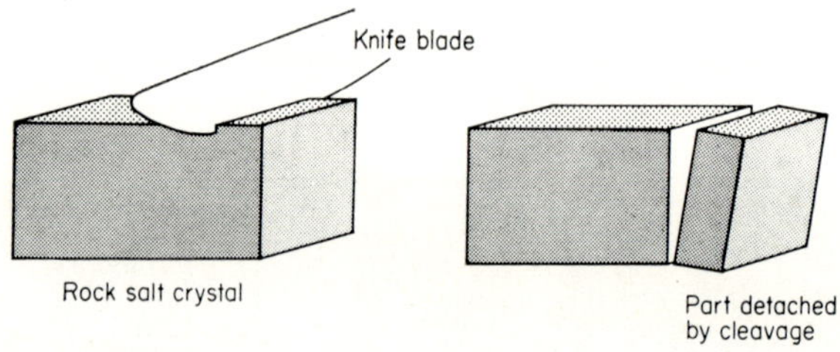

Fig. 3.2. (*a*) Rock salt crystal, with knife blade in contact; (*b*) cleavage fragment detached after sharp blow on back of knife blade.

large number of small rectangular parallelepipeds. In the same way, calcium carbonate can be broken into rhombohedra—parallelepipeds whose faces are congruent rhombuses and of which two of the opposite corners have edges that meet in exactly equal angles.

Sometimes a crystal has only one plane of cleavage, and it then breaks up into sheets. The outstanding example of such a material is the mineral mica; it breaks easily into transparent sheets of great uniformity, with only a few lines along which the thickness changes abruptly. If a good crystal can be obtained it can be cleaved into perfectly uniform sheets by the following method. A small incision is made in the side of the crystal and a piece of thin stiff card inserted; as this end is wedged in further the cleavage extends; and so long as the card is not allowed to reach the side no steps will be produced. When a large area of cleavage has been produced the two parts can be pulled apart. By repeating the operation, a thin perfectly uniform sheet results. (The work 'perfect' is seldom used by physicists; here it is justified because the cleavage occurs between atomic planes, as we shall explain in § 3.7.)

Sheets with areas of several square centimetres can be made in this way, and can thus be cut into small pieces of identical thickness. Such pieces have been used in certain optical experiments, as mentioned in § 10.5; it is difficult to see how these experiments could have been performed if mica had not existed.

3.2 *Crystallization*

Many theories have been put forward concerning the formation of crystals in mineral deposits. For example, diamonds are apparently the result of the crystallization of carbon from solution in molten minerals under conditions of high temperature and pressure. The large hexagonal crystals of basalt found at Giants' Causeway are thought to be the result of sudden cooling of the molten basalt rock.

In the laboratory, one method of producing crystals is by far the most popular: the material is dissolved in a suitable solvent which is then allowed to evaporate slowly. The material deposits when the solution becomes supersaturated and small single crystals form on the bottom of the containing vessel. These gradually enlarge, and if the process is slow enough a few large single crystals result. These should have good plane faces, but their bases take up the shape of the bottom of the vessel. To produce a perfect 'all-round' crystal, it is better to start with a seed crystal suspended by a thread in the middle of the solution. Good materials to work with are copper sulphate and alum (potassium aluminium sulphate) because they have high solubilities in water, a particularly simple solvent to use.

If a drop of hot saturated magnesium sulphate solution—Epsom salt—is placed on a cold microscope slide, one can actually see the crystals growing by means of a low-power microscope. The crystals are needle-like, and form beautifully straight rods before one's eyes.

Another process that is coming into favour, although it is more difficult to carry out, is to form crystals by deposition from the vapour. The material is heated in an enclosure until it volatilizes, and is then caused to condense on a cold point; a vapour will always condense on to the coldest point of its enclosure. If the rate of deposition is slow enough a single crystal should result.

The process of crystallization, either from a molten solid or from a suitable solution, is of immense importance in modern times. A method of separation of one compound from others in a mixture, known as fractional crystallization, is widely used by chemists today. The transistor industry depends upon the formation of single crystals, and even diamonds can now be made commercially for producing cutting tools (see p. 174). Crystal growth in all its aspects is at present a rapidly expanding subject.

3.3 *Theories of crystal formation*

Nicolaus Steno, a Danish physician, published in 1669 a treatise on his measurements of crystal of quartz. His main conclusion was that the angles between similar pairs of faces were always the same, whether the crystals were large or small. The extension of these measurements to other crystals, and the formation of a general law, were developed by Romé de l'Isle and the Abbé Haüy towards the end of the eighteenth century. Romé de l'Isle measured as accurately as he could the angles between the faces of crystals and established that for the same substances these angles are always the same although the shape may appear to be quite different. The Abbé Haüy supported these measurements and by studying the ways in which crystals could be easily broken or cleaved (§ 3.1) came to the conclusion that they were built up of small equal elements or bricks. Similar experiments can be tried by the reader using sugar lumps and a plane mirror. An octahedron can be formed by building a pyramid of sugar lumps on a plane mirror as shown in fig. 3.3. This construction, if observed from above the mirror so that

Fig. 3.3. Pyramid of sugar lumps on plane mirror, showing how crystal is built out of identical units.

the sugar-lump image can be seen, gives a realistic image of the octahedron, the shape of certain natural crystals. If instead of a single sugar-lump step for each layer other members are used such as two across and three up, then regular shapes can be obtained similar to the shapes of naturally occurring crystals. By this building method Haüy found that the secondary (or unusual) forms of a crystal were related to primitive form or 'cleavage nucleus' by the arrangement of the steps and that the width and height of a step are always in a simple ratio rarely greater than 1:6.

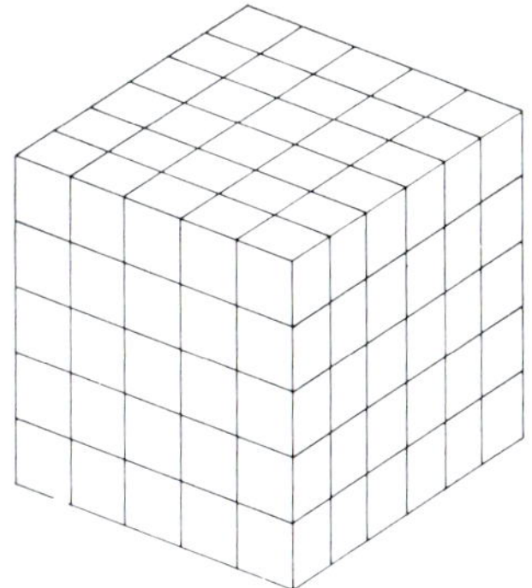

Fig. 3.4. Lattice derived from fig. 3.3 (not to scale).

The idea that crystals were composed of identical units, regularly stacked in three dimensions, became generally accepted at the beginning of the nineteenth century and was systematized by the introduction of some basic concepts. The most important was the *crystal lattice*. Let

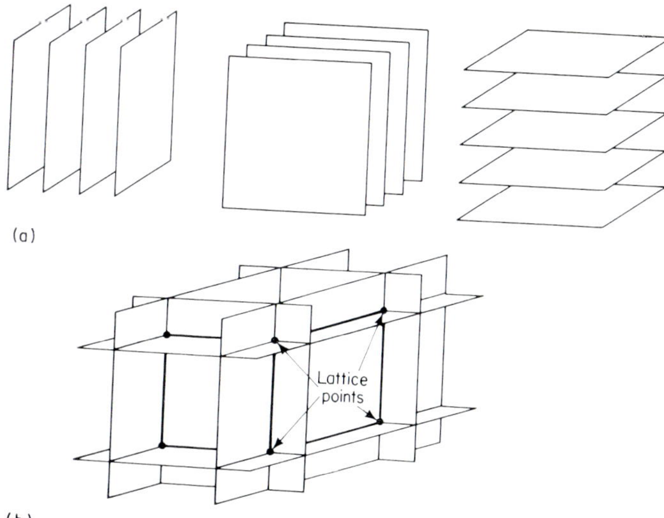

Fig. 3.5. (*a*) Three perpendicular sets of equally spaced planes; (*b*) superposition of sets of planes such as those shown in (*a*), with the resultant lattice points at the intersections.

us imagine that there is a point at the centre of each of the sugar lumps in fig. 3.3, and that these points remain, even if the lumps are taken away; these points have the property that, except for those in the faces, they all have exactly similar environments, similarly orientated (fig. 3.4). The complete set of points is the crystal lattice, which can be alternatively defined as the points of intersection of three—in general, unrelated in direction—sets of parallel equidistant planes (fig. 3.5).

In the example that we have used, of course, the three sets of planes are not unrelated; they are equally spaced and mutually at right angles. This produces what we term a *cubic lattice*, the conditions for which we shall discuss later. The general lattice shown in fig. 3.5 can be defined in terms of its unit cell, shown by the heavier lines in this figure; if we know the lengths of each side of the unit cell and their relative directions, we can construct the complete lattice. It should be noted that, in the general case, there is no absolute choice of unit cell; the same lattice can be described in terms of different units, as we can see in three dimensions in fig. 3.6.

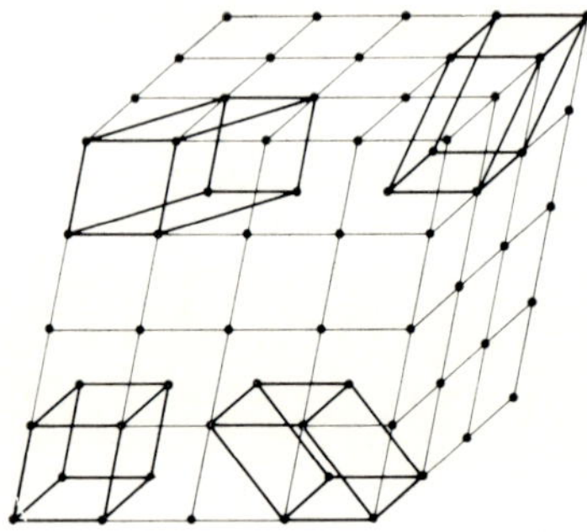

Fig. 3.6. Unit cells of different shapes in the same lattice.

The lattice has several interesting properties. We can draw lines in many different directions passing through rows of lattice points (fig. 3.7); we can draw planes, called *lattice planes*, that contain regularly arranged sets of points (fig. 3.7). Parallel to any one lattice plane we

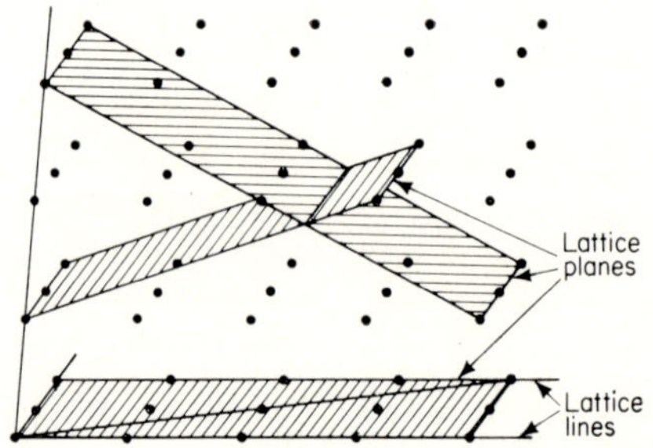

Fig. 3.7. Lattice lines or rows and lattice planes.

can draw sets of equally spaced planes (fig. 3.8) which contain *all* the points of the lattice; the separations of these planes are known as *planar spacings*. There is an infinite number of such sets of planes, those with large spacings being well populated with lattice points and those with small spacings being sparsely populated.

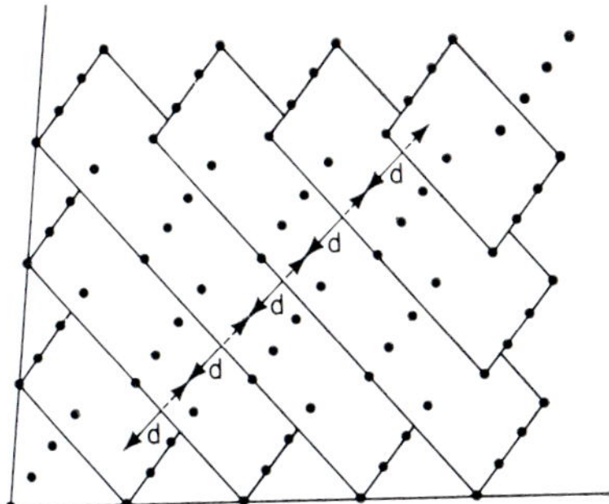

Fig. 3.8. A set of equidistant planes passing through all the lattice points shown in fig. 3.7.

In the nineteenth century these concepts were of little importance. The nature of the unit cell could be a matter of speculation only; was it some sort of container for atoms, was it a single molecule, or perhaps a single atom? None of these questions could be answered. Then suddenly, as we shall show in the next chapter, they sprang into importance when X-ray diffraction was discovered, and W. L. Bragg used them as a basis for his derivation of his now famous law.

3.4 *Miller indices*

The theoretical ideas introduced in the last section require a mathematical method of representing the various concepts introduced. In particular we have to find a way of representing the various faces that occur on a crystal. The method now in use was devised by W. H. Miller, Professor of Mineralogy in Cambridge from 1832 to 1881.

His method involved taking any three faces of a crystal, none of which is parallel to the other two, as planes of reference to define the other faces. Consider three planes intersecting at a point O, each plane being parallel to each of the three chosen faces of the crystal (fig. 3.9 *a*). The intersections of these planes with one another give the straight lines OX, OY and OZ. It is usually easy to choose for any particular crystal three planes which are clearly of importance; often two or more planes can be found which are mutually at right angles.

Any fourth plane of the crystal will cut the axes in points such as A, B and C in fig. 3.9 *a*, and the orientation of this plane is fixed if the lengths OA, OB and OC are known. In fact, since we are not concerned with the exact position of the face ABC the plane is defined by the ratios OA : OB : OC of the three intercepts.

Select any such plane and let the lengths of the intercepts be a, b and c respectively. The plane selected is called the *parametral plane* and the values of a, b, and c the *parameters*. It is possible with many crystals to select a plane which gives some simple relationship between the values of a, b and c. For example, for cubic crystals $a = b = c$.

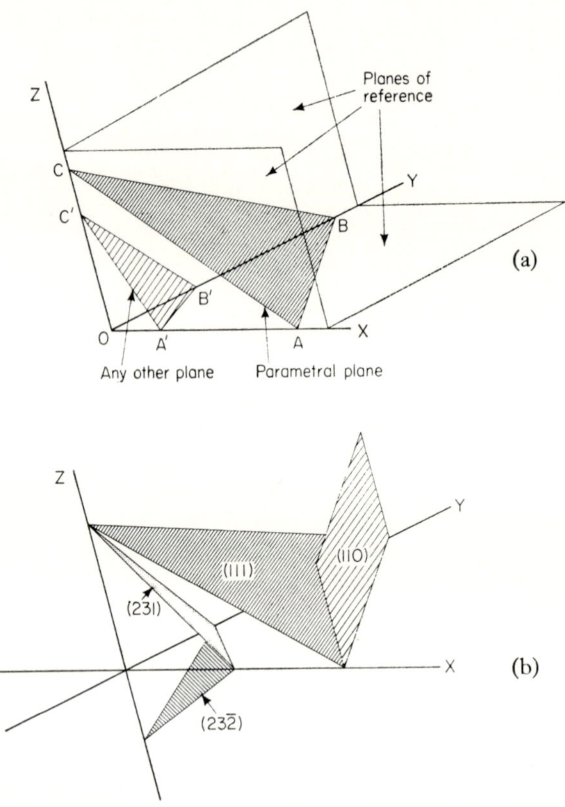

Fig. 3.9. (*a*) Set of axes—OX, OY and OZ—formed by the intersections of three chosen faces; (*b*) some planes of simple indices.

A plane parallel to another face of the crystal would cut the axes in three points A′, B′ and C′. The intercepts OA′, OB′ and OC′ are of course different from OA, OB and OC and may be written as $OA' = a/h$, $OB' = b/k$, $OC' = c/l$; if the parametral plane has been sensibly chosen it turns out that h, k and l are in the ratios of small integers numerically less than 6. This is known as the *law of rational indices*. The numbers h, k and l are called *Miller indices* and are enclosed in round brackets—(hkl).

Any of the indices can be positive, negative or zero. It will be zero if the intercept that the plane makes on the corresponding axis is infinite; that is, the plane is parallel to the axis (fig. 3.9 *b*). Thus, for example, all the faces with $l = 0$ form what is called a *zone* parallel to the *c* axis. If two of the indices are zero—and therefore the third is unity, since Miller indices do not have a common factor—the face is parallel to the plane containing two of the axes; such planes are of great importance. An index is negative if the plane makes a negative intercept on the corresponding axis (fig. 3.9 *b*). The minus sign is placed above the index as shown in the figure.

For a regular octahedron the faces are (111), $(\bar{1}11)$, $(1\bar{1}1)$, $(11\bar{1})$, $(1\bar{1}\bar{1})$, $(\bar{1}1\bar{1})$, $(\bar{1}\bar{1}1)$ and $(\bar{1}\bar{1}\bar{1})$. For a cube the faces are (100), (010), (001), $(\bar{1}00)$, $(0\bar{1}0)$ and $(00\bar{1})$.

The lines OX, OY and OZ are known as the *axes of* the crystal, and angles between them (YOZ $= \alpha$, ZOX $= \beta$ and XPY $= \gamma$) are of considerable importance. It is found that there are often relationships between the parameters *a*, *b* and *c* and between the angles α, β and γ, and on the basis of these relationships seven *crystal systems* can be defined as shown in the following table. Typical crystals belonging to these systems are shown in fig. 3.10.

System	Parameters	Angles
Cubic	$a = b = c$	$\alpha = \beta = \gamma = 90°$
Tetragonal	$a = b \neq c$	$\alpha = \beta = \gamma = 90°$
Trigonal	$a = b = c$	$\alpha = \beta = \gamma \neq 90°$
Hexagonal	$a = b = c$	$\alpha = \beta = 90°\ \gamma = 120°$
Orthorhombic	$a \neq b \neq c$	$\alpha = \beta = \gamma = 90°$
Monoclinic	$a \neq b \neq c$	$\alpha = \gamma = 90°\ \beta \neq 90°$
Triclinic	$a \neq b \neq c$	$\alpha \neq \beta \neq \gamma$

3.5 *Crystal symmetry*

We must now ask why these seven crystal systems exist. What is particular about the relationships between the parameters and the angles of the unit cell, and why are there only seven crystal systems? The answers lie in the concept of *symmetry*: crystals do not grow in a random way but obey rules which often lead to the specific relationships shown in the table. The complete basis of crystal symmetry requires a text book of its own, and we cannot do more here than summarize the conclusions.

The symmetries possible in a crystal are of three types, related to a point, a line or a plane. These *symmetry elements* are referred to as a centre of symmetry, an axis of symmetry and a plane of symmetry respectively.

A crystal has a centre of symmetry if the faces occur in pairs parallel to each other and on opposite sides of the crystal. The cube and the regular octahedron are obvious examples of this type of symmetry (fig. 3.11).

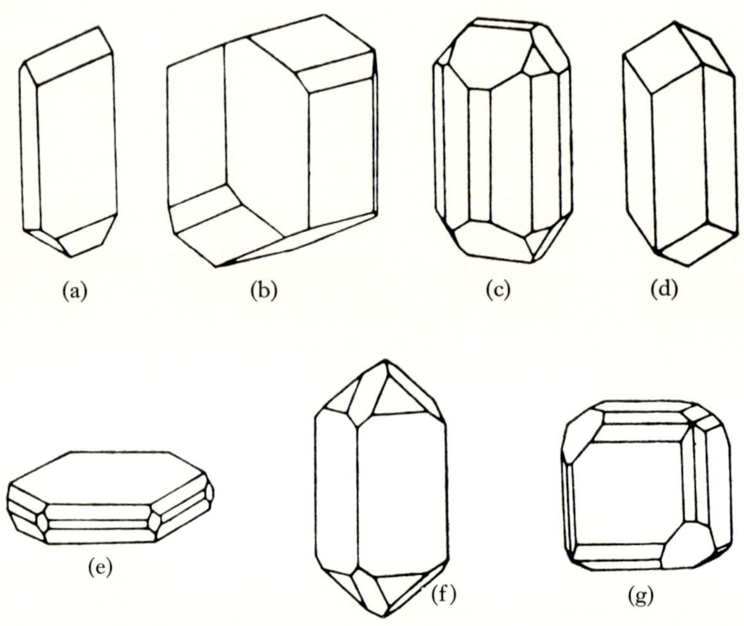

Fig. 3.10. Crystals belonging to the seven crystal systems. (*a*) Triclinic, strontium hydrogen tartrate; (*b*) monoclinic, borax; (*c*) orthorhombic, lead sulphate; (*d*) trigonal, calcite; (*e*) hexagonal, iodoform; (*f*) tetragonal, rutile, TiO_2; (*g*) cubic, sodium chlorate. (Courtesy of C. W. Bunn.)

A plane of symmetry is any plane surface which divides a crystal into two equal and similar halves, each of which is a mirror image of the other as shown in fig. 3.3. In the octahedron, fig. 3.11, any plane pass-

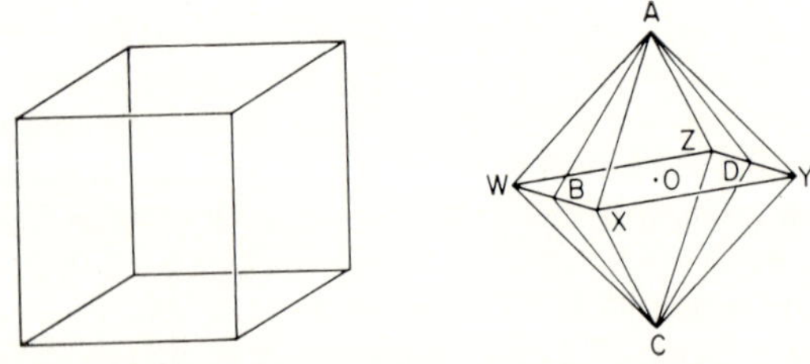

Fig. 3.11. Cube and regular octahedron as examples of shapes with centres of symmetry.

ing through four apices (such as the plane WXYZ) is a plane of symmetry. Also a plane passing through ABCD is a plane of symmetry.

A body possesses an axis of symmetry if rotation about a line turns a body in such a way that it appears indistinguishable from what it was to begin with. If it were not for the lettering on it, a pencil with hexagonal section will look just the same when it is rotated through 60°; it is said to have six-fold symmetry.

A regular octahedron and a cube have one centre of symmetry, thirteen axes of symmetry (of three kinds) and nine planes of symmetry. This number of elements of symmetry is the highest in the whole crystalline scheme.

From the description of the symmetry elements we can see that the centre of symmetry and the mirror plane are unique as types of symmetry element, but the rotation axis appears to have an infinite number of possibilities. For example, the 50p coin has seven-fold symmetry,

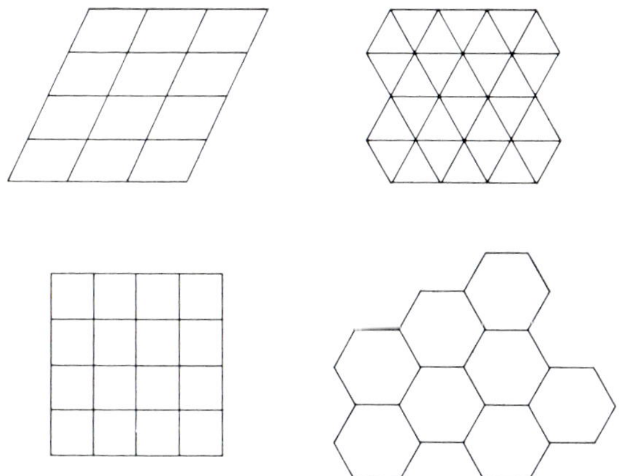

Fig. 3.12. Filling two-dimensional space with parallelograms, equilateral triangles, squares and regular hexagons.

but this does not occur in crystals. Why? The answer is that it is impossible to make a compact repeating pattern with this shape. To show this practically, take about twenty 50p coins and try to pack them closely together; you will find that there are always some spaces left. (If this experiment is too expensive, borrow one, trace its shape on cardboard, and cut out about twenty!)

The only figures that can be used are the parallelogram (2-fold symmetry), the equilateral triangle (3-fold), the square (4-fold) and the hexagon (6-fold); these are shown in fig. 3.12. These are the only axial symmetries that are found in crystals, but it is convenient to add another—the one-fold axis, which really means the absence of symmetry.

The seven crystal systems result from different combinations of these rotation axes, centres of symmetry and mirror planes, ranging from no symmetry at all to the full complement of elements in the cubic system. The lattice must conform to the symmetry of the crystal. Every lattice is centrosymmetric, and therefore the most general lattice—triclinic—has either no symmetry or a centre of symmetry. The monoclinic lattice, which has one axis perpendicular to the other two, has either a plane of symmetry or a two-fold axis, or it can have both. The triclinic system is thus said to have two subdivisions, called *point groups* or *crystal classes*; the monoclinic system has three point groups. There are thirty-two altogether in the seven crystal systems.

One further complication is that symmetry may not simply relate one part of a unit of pattern to another part of the same pattern; it may relate two parts of units in different unit cells. The corresponding symmetry elements—called *glide planes* and *screw axes*—involve translation as well as reflection or rotation, but we shall not consider them further here; they extend the number of possible symmetry combinations to 230, known as *space groups*.

The point groups of a crystal can be determined from measurements of crystal faces, but space groups can be determined only with the help of X-ray diffraction. It is remarkable therefore that the whole of space-group theory was worked out independently by three men—Fedorov in Russia, Schoenflies in Germany and Barlow in England—in the closing years of the nineteenth century, before there was any possibility of verifying it experimentally. Their work was invaluable once X-ray diffraction got under way.

3.6 *Significance in terms of atomic arrangement*

Robert Hooke in 1665, remarking upon the regularity of the small crystals of quartz found inside the cavities in flints, suggested that the crystals were built up of spheroids. About the same time Huygens in Holland was studying crystals of calcite (Iceland Spar), noting the double refraction and the very clean cleavage planes (§ 3.1). Huygens agreed with Hooke on the theory of the structure of crystals. The very thin sheets of mica which can be obtained by cleavage (§ 3.1) are very good examples of the phenomena which indicate that the structure of a crystal is built up of a regular symmetrical pattern of points. It seems logical to suppose that the crystal is held together by attractive forces between the points and that cleavage takes place along the planes where this force is weakest. Since the highly symmetrical shape of a crystal requires a highly symmetrical arrangement of the points forming the crystal, it is not surprising that cleavage planes should be so remarkably plane and angularly exact.

We now know what the crystal units are. They are atoms or combinations of atoms. These combinations may be single molecules or groups of small numbers of molecules; they may be pairs of oppositely charged ions, which themselves may be single atoms or groups of different atoms.

An infinite variety of possibilities exists, and Nature has devised an extraordinarily large number of ways of arranging in crystals the small number of different atoms that exist.

If the unit is a single atom its environment is necessarily highly symmetrical, which explains why so many elements have structures of high symmetry. If the unit is a molecule, the arrangement need not be symmetrical; hexamethylbenzene, with one molecule in the unit cell, is triclinic, for example. If the unit cell contains two or four molecules they are normally arranged in a symmetrical way, often with screw axes or glide planes (§ 3.5), which produces the symmetry observed in the crystal. The vast majority of organic crystals are monoclinic, and nearly all the rest are orthorhombic; high symmetry is spurned by organic compounds is general.

Ions and combinations of ions are not so easy to generalize about. Inorganic salts, often with the aid of water of crystallization, form crystals of all sorts of symmetry. We shall discuss these matters in more detail in Chapter 9, when we consider the information that crystal structures have provided about atomic arrangements in crystals.

3.7 *Physical properties and symmetry*

When we wish to know the value of a physical property of a substance, we can usually find it in some book of tables; we rarely ask whether it is reasonable to expect that the physical property has a specific value. In fact, however, the expectation is justified only because most solids, as we use them normally, are made up of random arrangements of single crystals, and the values given in the tables are averages. If we were dealing with a single crystal, much more than one value might be needed to specify the property exactly. Let us see how this complication arises.

We shall choose four properties—density, thermal expansion, Young's modulus and refractive index—because these illustrate the variety of complications that can exist, in order of increasing complexity. On a *macroscopic* scale—that is, on a scale on which atoms are not separately observable (in contrast to the *microscopic* atomic scale)—density is constant from point to point in a crystal because all unit cells are the same. In other words, a crystal is *homogeneous*. Only one value is needed, whether for single crystals or polycrystalline matter.

Thermal expansion is different. As we can see from, for example, fig. 6.4, in different directions we meet different sequences of atoms, and when a crystal expands there is no reason why the different directions should expand equally. We may express this fact by drawing a set of vectors from an arbitrary point, the direction representing the direction in the crystal and the length representing the coefficient of expansion. It can be shown theoretically that the three-dimensional figure so produced is a *triaxial ellipsoid*—a body (fig. 3.13) of which all the sections through the centre are ellipses, which can be circles in special cases. Such a body has three planes of symmetry, and the lines perpendicular to these are the three axes—principal axes—which give the figure its name. The length of any radius vector of this figure is a measure of the thermal expansion of the crystal in the corresponding direction.

Now the figure must necessarily have the symmetry of the crystal. Thus for a cubic crystal, since the three axes of the ellipsoid are equal,

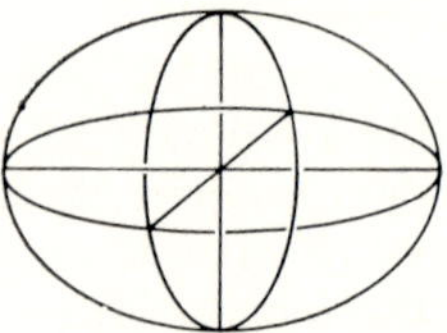

Fig. 3.13. Triaxial ellipsoid, representing a simple physical property of an anisotropic crystal.

the figure must be a sphere, which is a special case of an ellipsoid. Therefore the coefficient of thermal expansion is independent of direction. Cubic crystals are said to be *isotropic*. For crystals that are tetragonal, trigonal or hexagonal, two of the axes must be equal in order to preserve the symmetry, and therefore one of the principal sections is

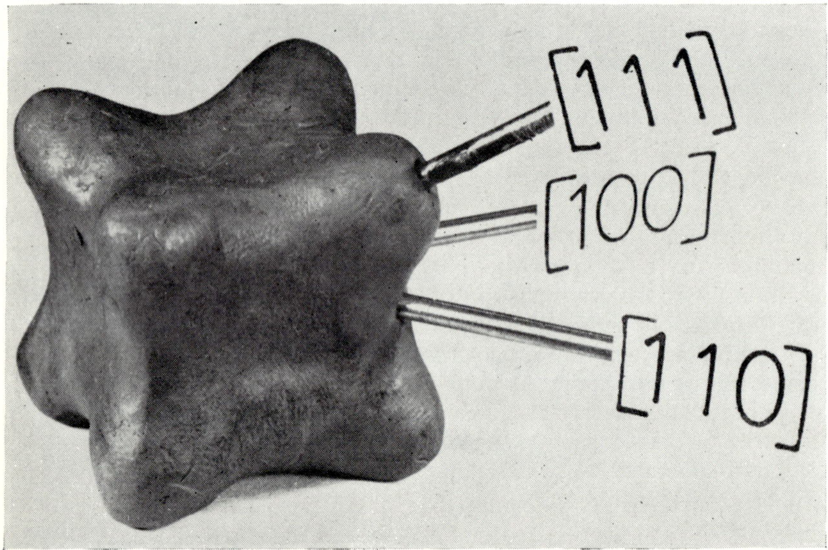

Fig. 3.14. Figure representing a possible variation with direction of Young's modulus in an isotropic crystal.

circular; the axis perpendicular to it is different in length, and the crystals are therefore said to be *uniaxial*. For crystals of lower symmetry, the figure is a general triaxial ellipsoid; this can be shown to have two circular sections and the lines perpendicular to these give the name *biaxial* to crystals with these symmetries. All crystals other than cubic are said to be *anisotropic*.

But cubic crystals are not isotropic in all their properties. More complicated effects arise in elastic deformation because, for example, when a tension is applied to a piece of matter it does not produce an effect only in the direction of application; there is a transverse effect—the Poisson contraction—as well. For this reason the figure for Young's modulus is more complicated than a triaxial ellipsoid, and, although the figure must conform to the symmetry of the system, it can have odd bumps and depressions in different directions (fig. 3.14). (Indices in square brackets [hkl] indicate directions.) In other words, Young's modulus may vary with direction even for cubic crystals.

The most exciting results appear, however, for light. A single refractive index for monochromatic light is appropriate only for an unusual substance such as glass or for cubic crystals; for all others the behaviour of light needs not one surface to represent it, but two. At first sight this may seem odd: when an electromagnetic wave passes through a crystal the electrons are displaced in the direction of the electric vector; there are no transverse effects as there are for deformation. Complications arise, however, because the behaviour of the light depends upon its plane of polarization.

A beam of light is said to be plane-polarized when the electric vector of the electromagnetic wave is confined to a particular plane; the magnetic vector is confined to a plane at right angles. Ordinary light can be considered to be a mixture of plane-polarized waves whose planes are randomly distributed. Let us imagine a beam of plane-polarized light to fall on a uniaxial crystal along the *optic axis*—the axis of symmetry. Clearly, because of symmetry, the light will behave in the same way whatever the plane of polarization, and so no unusual effects are observed; its speed, and consequently the refractive index, is independent of the direction of the plane.

But now consider what happens if the light passes perpendicularly to the optic axis. If the plane of polarization is parallel to the optic axis, again no unusual effects are observed, because the electric vector is meeting the same groups of atoms similarly arranged. But, if the plane of polarization is perpendicular to the optic axis, the speed is dependent upon the direction of the plane and if we represent the speed as a vector, as we did for thermal expansion, we obtain the same sort of figure.

The strangest result of all appears when we consider light polarized in some intermediate plane; we should expect some intermediate properties. This is fact does not happen; the light divides into two parts, one polarized in the plane containing the optic axis, the other polarized at right angles. These travel with different speeds and so give two different refractions (fig. 3.15). The effect is known as *double refraction*. For one of the beams, called the *ordinary beam*, the refractive index is constant and the light behaves as though the crystal were isotropic. For the other, called the *extraordinary beam*, it behaves as though the crystal is anisotropic (fig. 3.16). For biaxial crystals, the light splits into two extraordinary beams.

Fig. 3.15. Double refraction in a crystal of calcite.

This is one of the most fascinating effects in physics. It was discovered by Bartolinus in 1669, and was investigated by Huygens who gave the full theory of it although he did not know what sort of waves were involved; in fact his great contemporary, Newton, thought that double refraction could be explained only if light were particles like little magnets—hence the name polarization.

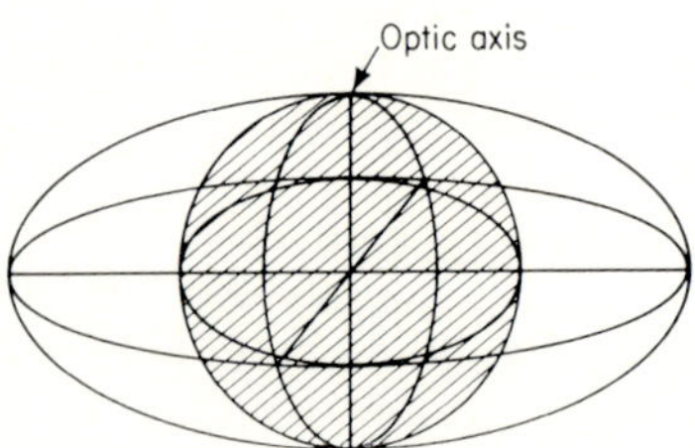

Fig. 3.16. The behaviour of light passing through a uniaxial crystal. The shaded part represents a sphere, included within an ellipsoid of revolution.

There are, of course, other properties of crystals that we could discuss, but the four that we have included show all the variations that are possible in the different crystal systems. As we have said, the optical effects are the most interesting and informative, but it is nevertheless fortunate that there are no corresponding effects with X-rays; the refractive index of a crystal for X-rays is so near to unity that no variation with plane of polarization is detectable. The interaction of X-rays with crystals is complicated enough without our having to tackle optical anisotropy as well!

3.8 *Isomorphism*

There is one further concept in crystallography that has since proved to be of great practical value. This is *isomorphism*, first discussed by Mitscherlisch in 1819, when the concept of atoms had only just been accepted and atomic and molecular weights were being determined for the first time. Mitscherlisch stated that if compounds crystallized in similar forms, they probably had their constituent atoms combined in the same proportions; if the chemical formula of one compound were known, the others could then be deduced. For example, since sodium chloride is cubic and has the formula NaCl, potassium chloride, which is also cubic, should have the formula KCl. Such crystals are said to be isomorphous.

The most remarkable series of isomorphous compounds is undoubtedly the alums, double sulphates of a monovalent atom and a trivalent atom, the typical formula being $K_2SO_4.Al_2(SO_4)_3.24H_2O$ or $KAl(SO_4)_2.12H_2O$ for brevity. This substance crystallizes as beautiful large octahedra, and so do many double sulphates containing other monovalent atoms or ions, such as NH_4, and the trivalent atoms Fe and Cr; also the sulphate ions can be replaced by selenate ions. If the formula of one is known, then the others all follow, and this rule gave considerable help to the early 'atomic' chemists in deciding an unknown chemical formula.

We now know that cubic crystals do not provide good tests of isomorphism; crystals can be cubic and yet have different crystal structures, as we shall show later (p. 132) for NaCl and CsCl. The alums also do not all have similar atomic arrangements (§ 8.5). But when we find that copper sulphate and copper selenate, which are triclinic, crystallize with similar unit cells (§ 8·5) we know that their atomic arrangements must be almost identical.

Since a pair of isomorphous crystals must have all except the replaceable atoms in identical arrangements, we have an extremely useful 'variable' to assist in determining crystal structures, as we shall show later in Chapter 8.

CHAPTER 4

diffraction of X-rays

4.1 *Laue's treatment*

We have seen in Chapter 2 how the diffraction of X-rays was discovered, and how beautifully symmetrical photographs were soon obtained (fig. 2.5). But these were only the beginning of the story; a complete explanation of the patterns had to be obtained and this proved to be extremely difficult. Laue set himself the task of devising a theory of the diffraction of a monochromatic radiation by a three-dimensional diffraction grating.

His theory was based upon that for the ordinary diffraction grating; if radiation of wavelength λ falls normally on a grating of spacing d, orders of diffraction are formed in accordance with the equation:

$$n\lambda = d \sin \theta_n, \tag{4.1}$$

where n is an integer and θ_n is the angle at which the nth order is formed (fig. 4.1). For this diffracted beam, the path difference between the successive scattered waves is $n\lambda$.

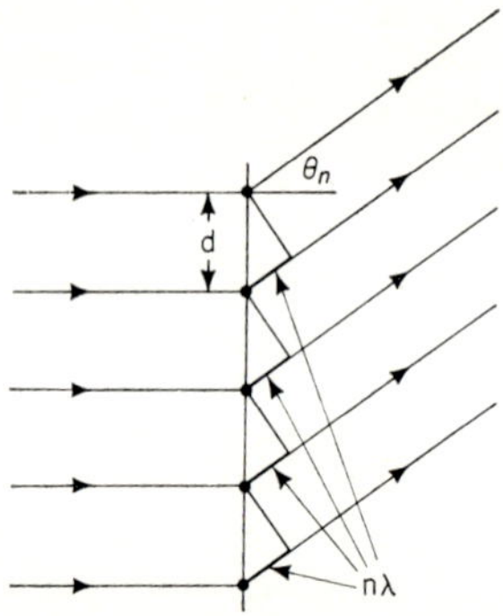

Fig. 4.1. Production of nth order of diffraction from a grating of spacing d, normally illuminated. The successive path differences are $n\lambda$.

Now, for a crystal there is no simple quantity corresponding to d; there is a unit cell with three edges a, b and c (p. 34). There is nothing to correspond to the plane of the grating, which acts as a reference for measuring θ, and thus Laue's theory had to be much more general. He produced three equations corresponding to equation (4.1), each of the equations being associated with one of the quantities a, b and c. The three equations had to be satisfied simultaneously if an order of diffraction were to be produced.

This was the difficulty: it was unlikely that all three equations would be satisfied at once, and Laue had to postulate certain relations between the lattice constants of ZnS (p. 20) and the wavelength to explain any of the spots at all. The fact that the unit cell was a cube made the theory simpler; there was only one lattice constant to deal with, and values of λ/a could be specified.

Only a few of the spots produced by the ZnS crystal could be explained in this way. Laue was thus forced to assume that more than one wavelength was present as already mentioned in § 2.7, so that more of the spots could be accounted for. But always there were some spots that did not fit in, and the artificiality of the procedure soon became obvious. The theory was clearly right, but it did not explain the facts. What was wrong?

4.2 *The Braggs*

The answer to the question came through the ideas of two Englishmen, W. H. and W. L. Bragg, both of whom later became noted for their ability to see uncomplicated routes to the solution of physical problems. Together they formed what is probably the most successful father–son combination in the whole of physics, and a brief account of their history is worth telling; in some ways it shows that science is not always carried out in the coldly impersonal way that many people think.

W. H. Bragg had not at first intended to become a research scientist. He was a bright boy at school and went to Cambridge to read mathematics, graduating exceptionally well in 1884. On the basis of this degree he worked up sufficient courage to apply for a professorship in Adelaide in 1886, when he was only twenty-four, and to his surprise he was successful. The chair was that of Mathematics and Physics, and he is reported to have said that he learned his physics on the sea voyage out; fortunately there were no aeroplanes then to shorten the journey!

In Adelaide he took his teaching duties extremely seriously and trained himself to become an excellent lecturer. (He ultimately became one of the best scientific lecturers in the world.) He did not, however, consider original research to be part of his duties. In 1904, however, when he was forty-two, the Australian Society for the Advancement of Science met in Adelaide, and he was invited to become its President. He thought that he ought to give an address on some research topic, and chose radioactivity, one of the current scientific mysteries. Within four years he had become one of the foremost authorities on the subject of the absorption of α-particles and on the strength of his reputation he was invited to return to England to become Professor of Physics at Leeds.

In his research he also investigated the ionization produced when X-rays passed through a gas, and came to the conclusion that X-rays must be particles something like uncharged α-particles—'an electron which has assumed a cloak of darkness in the form of sufficient positive electricity to neutralize its charge.' In this idea he was supported by many other people.

Then came the experiments of Laue, Friedrich and Knipping. Bragg was puzzled, but was more ready to compromise than some of his supporters were; he suggested that the problem was not to find whether the particle or wave theory was correct, but to find another theory which would possess the capacity of both.

His son, W. L. Bragg, had the advantage of living in a scientific atmosphere, mixed with the outdoor life for which Australia is famous. His playthings were scientific apparatus that he put together for himself—simple electric motors that worked, Morse tapping keys, telephone

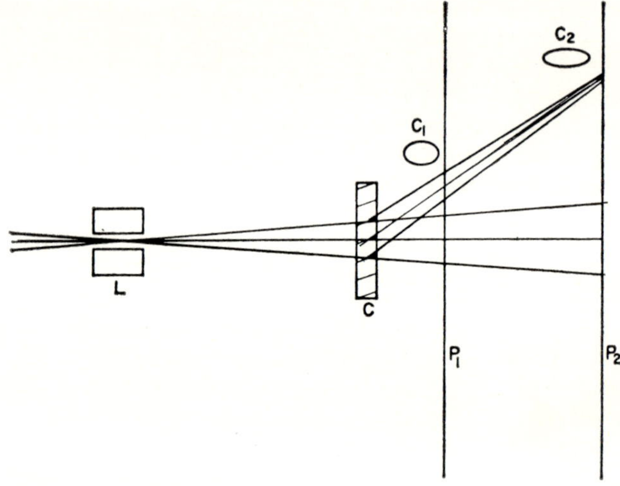

Fig. 4.2. W. L. Bragg's original drawing showing how Laue spots change shape as the distance of the photographic plate from the crystal changes.

sets and so on. He did well at school and when the family moved to Leeds he went to study physics at Cambridge. In that great year, 1912, he was a research student working under J. J. Thomson on the mobility of ions in gases. But he was a keen supporter of his father, and could not keep his mind off the X-ray problem; he wondered if X-ray 'particles' were somehow passing through channels in the crystal to produce what were thought to be diffraction patterns. But the beautiful symmetry of the photographs of ZnS convinced him that the patterns could not be explained in this way.

J. J. Thomson must have been a liberal supervisor; he allowed his young research student to be distracted from his chosen topic in a way that few supervisors would approve of nowadays. For a brilliant idea had struck W. L. Bragg; he claims that he knew exactly the place in the

'Backs' at Cambridge where it came to him. Were the X-rays being reflected from mirrors in the crystal? He tried, with J. J.'s encouragement, to reflect X-rays from mica, which has a beautiful plane of cleavage (§ 3.1); a very strong spot was obtained in the right place on a photographic plate. He tried transmitting the rays through a crystal of ZnS, varying the inclination of the crystal and the distance of the photographic plate; the position of the spots and their *shapes* were consistent with the reflection of a beam of circular section from a plane mirror (fig. 4.2). At the age of twenty-two he had hit upon one of the most prolific ideas in physics.

4.3 *Bragg's equation*

But what were the reflectors in the crystal? Could they be the lattice planes (§ 3.3)? Clearly, one plane would be too weak, but could they combine together to produce the strong reflections observed?

The answer to this question now seems obvious, but it was not so in 1912; there were no parallels to draw upon. The conditions must be

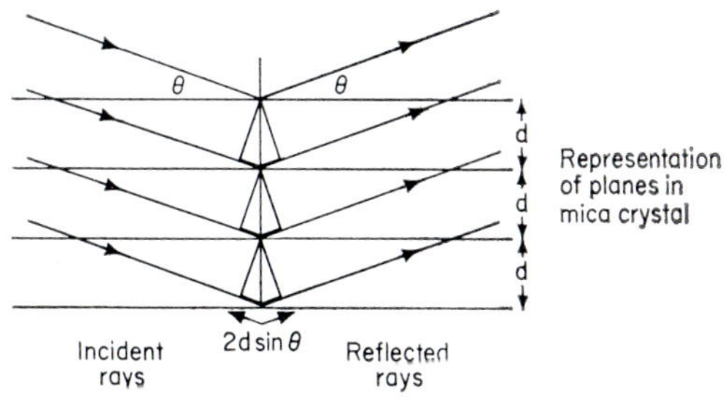

Fig. 4.3. Reinforcement of waves reflected from successive planes of a crystal.

such that the waves from successive planes reinforced each other. If a plate of mica is set obliquely in the path of a narrow X-ray beam of fixed wavelength (fig. 4.3), each plane parallel to the surface will reflect the beam; but only in certain orientations will the waves from successive planes reinforce each other. The condition that this reinforcement should take place is simply

$$n\lambda = 2d \sin \theta, \tag{4.2}$$

where d is the distance between successive planes. This is the equation that set the subject on its feet; it is known as *Bragg's equation*, and the values of θ that satisfy it, for different values of the integer n, are called *Bragg angles*.

It must be emphasized that the equation is deceptively simple. The important idea in it is that of reflection of the waves from planes, but it should be realized that waves can be scattered from planes in other directions also if the planes contain regular arrangements of atoms. The general equation, as we can see from fig. 4.4, is:

$$n\lambda = a(\cos\theta - \cos\phi), \tag{4.3}$$

where θ is the grazing angle of incidence, ϕ is the grazing angle of scattering, and a is the separation of the atoms. Bragg's idea was to con-

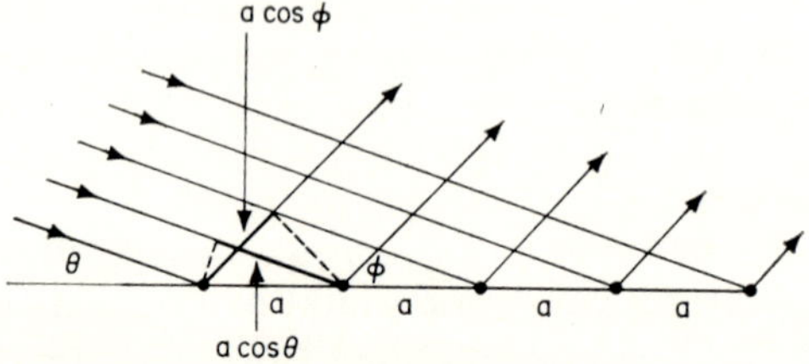

Fig. 4.4. Derivation of the general equation for a diffraction grating.

sider only the one solution, $\theta = \phi$, for this equation, and to suggest that any other solution could be regarded as reflection from some other set of lattice planes in the crystal.

Another deceptive property of Bragg's equation is its similarity to equation (4.1). Why does Bragg's equation have θ as the grazing angle of incidence instead of the inclination to the normal, and where does the factor 2 come from? The answer to both these questions comes from diffraction-grating theory; the plane of the grating has nothing to do with a lattice plane; it is the spacing—which is perpendicular to the lattice planes—that matters (fig. 4.5). Thus Bragg's grazing angle of incidence should be regarded as the angle between the incident ray and the *normal* to the spacing.

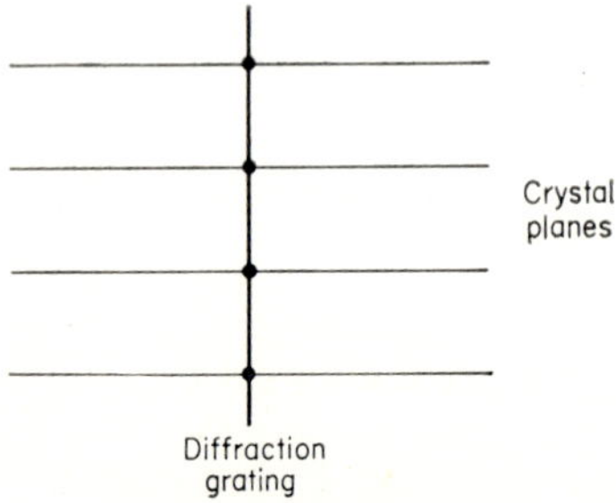

Fig. 4.5. Relation between a plane diffraction grating and the lattice planes of a crystal grating.

The factor 2 enters because the grating is not being used with normal incidence, which is the condition for validity of equation (4.1); it is being used in an orientation that gives minimum deviation. It is not always realized that a diffraction grating can be orientated to give minimum deviation for each order of diffraction and that this method has certain advantages over the usual procedure.

4.4 *Development of the theory*

Although Bragg's equation was necessary for setting the subject on its feet, it can now be replaced by a more general treatment. Bragg's idea was to reduce the three-dimensional problem of working out crystal structures to sets of one-dimensional problems. In this way, as we shall show in Chapter 6, he was able to derive atomic arrangements in crystals when nothing at all was known about them. The sizes of atoms—absolute or relative—were unknown and the nature of chemical bonding was a subject of intense speculation.

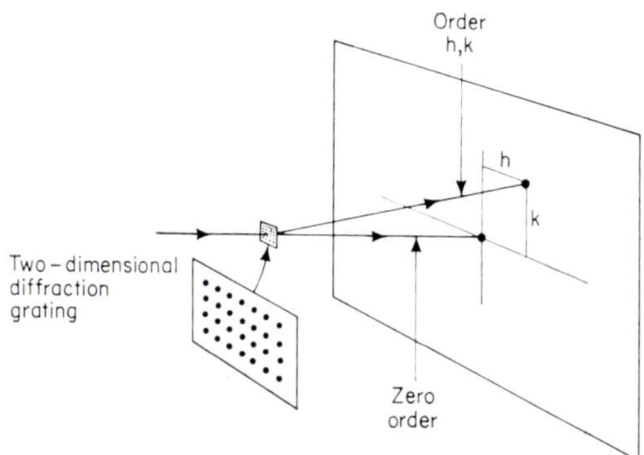

Fig. 4.6. Diffraction from a two-dimensional grating of holes, showing how an order is specified by two quantities h and k.

Laue's ideas, as we have seen, were not able to cope with these difficulties. But now that methods have become more general, and crystal-structure problems of immense complexity are now being successfully solved, Bragg's one-dimensional methods are quite inappropriate, and it has been necessary to revive Laue's theory. This theory must not however be regarded simply as applied mathematics; it has a sound physical basis that can be explained in terms of diffraction-grating theory, as we shall now show.

The quantity n in equation (4.1) is the number of wavelengths difference between the paths of waves scattered by successive elements of the grating. This number is all that is needed to specify the path difference. Let us extend this idea to a two-dimensional grating, which we can regard as composed of a set of holes arranged on a two-dimensional lattice (fig. 4.6). If we allow a beam of light to fall normally on this grating a set of orders of diffraction will be observed. But each order now has a direction in space that has to be specified by two angular components and therefore *two* integers are involved. Let us call them *h* and *k*. For an order of diffraction produced in the plane defined by the incident beam and one of the lattice axes, the corresponding integer will be zero. In general, the integers *h* and *k* may assume any values—positive, zero or negative—but usually, as for ordinary diffraction gratings, they will be small.

What happens when we extend these ideas to a three-dimensional grating? It is simple enough to say that now a third integer, *l*, is involved so that each order of diffraction is specified by three integers—*h*, *k* and *l*. But it is not so easy to see how they arise. In fact, the change from two to three dimensions is one of some complexity. Each order of diffraction is still defined by two angular components, but a third condition is necessary to decide whether the diffraction beam will be produced at all. If this condition is not obeyed, the order of diffraction will not be produced; only by varying the orientation of the crystal can the condition be satisfied.

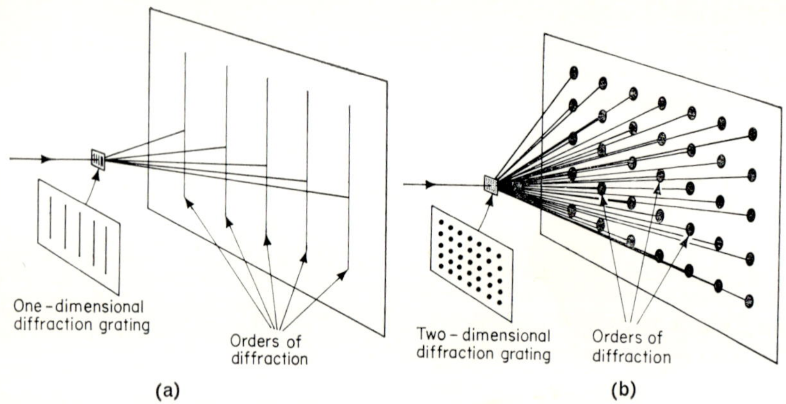

Fig. 4.7. Comparison of diffraction by a one-dimensional grating (*a*) and a two-dimensional grating (*b*). The arrays of orders are idealized; they would not in practice be exactly straight.

The sequence of events in one-, two- and three-dimensional diffraction is as follows. For a one-dimensional grating all the orders of diffraction are produced together, whatever the direction of incidence of the beam of light; each order is specified by a single angle (fig. 4.7 *a*). For a two-dimensional grating, all the orders of diffraction are again produced simultaneously, but they are now discrete beams, specified by two angular components (fig. 4.7 *b*). For a three-dimensional grating, the orders are produced separately; with an exactly parallel beam, no orders are produced, since the angle of incidence is precisely fixed and it is unlikely to satisfy Bragg's equation: but as the orientation of the grating is varied, orders will flash out as the third condition is obeyed. The three integers—*h*, *k* and *l*—can be regarded as occurring because the grating is three-dimensional.

This approach seems to have little relation to that of Bragg. In fact it turns out, as it should do, that there is an exact relation. The three integers—h, k and l—are those that specify the directions of the reflecting planes for the given order of diffraction; these are like the Miller indices introduced on p. 33. But they convey more information than the Miller indices, which, as we saw, do not have a common factor. The indices of an order of diffraction do not have this limitation; they may have a common factor, and this is the integer n in Bragg's equation. Thus the first-order reflection from the planes (110) is given the symbol 110, the second-order reflection is 220 and so on. The order of reflection 633 can be regarded as the third order from the planes (211). In the early days, reflections were written in some such way—for example, 211(3). It can readily be appreciated that the symbol 633 is much neater as well as being more physically significant, since it specifies the order of diffraction.

4.5 *The reciprocal lattice*

The idea of representing each order of diffraction by three numbers suggests that there must be a three-dimensional way of representing the diffraction pattern of a crystal, the quantities h, k, l being the coordinates. This is indeed true, and the representation is another lattice which has a simple geometrical relation to the crystal lattice; it is called the *reciprocal lattice* of the crystal. It was first put forward by Ewald as a mathematical concept, useful for dealing with the interpretation of diffraction patterns. We shall show that it also has an important physical significance.

This can be understood by considering the ordinary diffraction grating. The equation (4.1) can be written as:

$$\sin\theta = n(\lambda/d); \tag{4.4}$$

the values of $\sin\theta$ at which the orders of diffractions occur lie at equally spaced intervals. The direct beam can be regarded as the zero order and so, since there are orders of diffraction at each side of zero, we may say that they can assume any integral value—positive, negative or zero—consistent with the limit set by the fact that $\sin\theta$ is less than unity.

If we compare gratings with different spacings, we find that the separations of the orders of diffraction are inversely proportional to d. This is the reason why the name 'reciprocal lattice' is used. A grating with a large spacing has a fine reciprocal lattice and one with a small spacing has a coarse reciprocal lattice.

Let us now consider a two-dimensional grating, which we can imagine as produced by the superposition of two gratings of slits (fig. 4.8 *a*). Each grating produces its own conditions for diffraction, which we can represent by its reciprocal lattice; the complete diffraction pattern will be strong only where both conditions are satisfied together—that is, at the intersections of the two sets of lines (fig. 4.8 *b*).

There are some obvious properties of this two-dimensional reciprocal lattice. First, its axes are perpendicular to the directions of the slits of the two gratings of which the diffracting lattice is made; in other words, the reciprocal axes, which we may call x^* and y^*, are perpendicular to the axes of the grating, y and x. Secondly, the unit cell of the reciprocal lattice, with edges a^* and b^*, is reciprocally related in shape to the unit cell of the grating, as we can see from fig. 4.8

We must now extend this idea to three dimensions. This is not so easy; the retina of the eye is two-dimensional, and, although the brain has found a way—by the stereoscopic action of a pair of eyes—to appreciate three-dimensional objects directly, there is no way for it to envisage three-dimensional diffraction patterns. As in other branches of physics, when we are faced with the necessity of exceeding the potentialities of the brain—as for example when we wish to explore a space of more than three dimensions—we have to resort to mathematics. For this reason, the three-dimensional reciprocal lattice is usually regarded simply as a mathematical figment. Essentially, however, it must be regarded as the diffraction pattern of the crystal lattice.

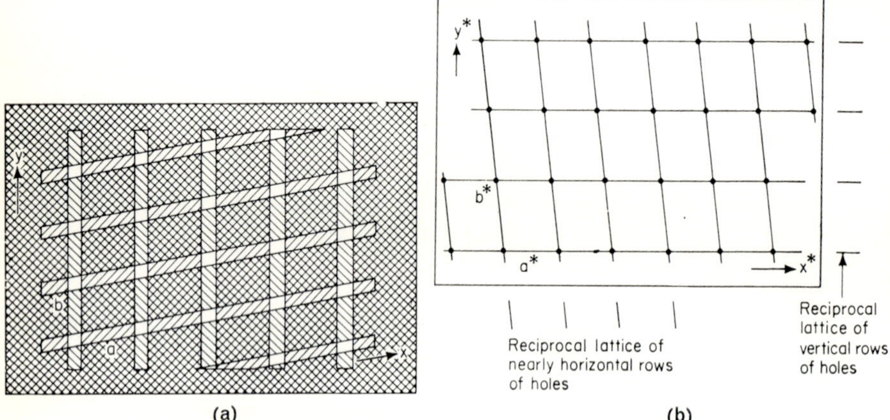

Fig. 4.8. (*a*) A two-dimensional grating of holes considered as a superposition of two gratings of slits; (*b*) the reciprocal lattice resulting from the grating shown in (*a*).

4.6 *Importance of monochromatic radiation*

We can now see the importance of the discovery of the characteristic radiations discussed in Chapter 2. The reciprocal lattice of a crystal is a collection of points only if a single wavelength is used for diffraction; if a range of wavelengths is used, each point is spaced out into a spectrum. One can appreciate the complexity this caused. The spectra, particularly in the higher orders, would overlap each other, and no clear-cut diffraction patterns would be observed. This was essentially the reason for Laue's problem (p. 20) in interpreting his first diffraction pattern: the spots that he obtained did not correspond to reciprocal-lattice points; they were points on the streaks caused by the continuous radiation (p. 22) emitted from the X-ray tube. Thus the spots—except for the occasional coincidence—were produced by different wavelengths and often represented points on several different streaks at once. This subject will be discussed in more detail in the next chapter.

4.7 *The physics of diffraction*

The lattice of a crystal is essentially an abstract concept; it does not exist except as a basis for constructing crystal structures. When we have talked about diffraction by a lattice we have been careful not to mention what radiation it was diffracting; in fact it could not diffract a radiation that had a physical existence. We must therefore ask what exactly does diffract X-rays.

To say that atoms diffract X-rays is not enough; we must know which particular property of the atoms is responsible. Now, X-rays are an electromagnetic radiation; that is, at any point in an X-ray beam—or in light or in radio waves, for that matter—there are electric and magnetic intensities operating at right angles and oscillating at high frequency. The frequency is very high—about 10^{18} Hz. If we could oscillate a magnet or an electrically charged body at this frequency, it would emit X-rays.

The electric field is capable of causing electrically charged bodies to move with the same frequency. Now atoms contain such bodies—electrons and protons. The proton is much more massive than the electron and therefore does not respond as actively; the electrons vibrate with the frequency of the X-rays and so become sources of X-rays. This is what we observe as scattering.

The waves that are scattered are closely related to the exciting wave of incident radiation. It is referred to as *coherent scattering* and is responsible for all the effects of X-ray diffraction that we shall describe in this book. Because large numbers of atoms are affected in a similar way, the diffraction spots that we have described are produced. If the atoms were not similarly excited, we should observe only a general blur; the scattering would then be said to be *incoherent.*

Since the scattering of X-rays is caused by electrons, we should expect that the heavier atoms, because they contain more electrons, would scatter better than lighter ones. This is in fact true, but the scattering is not proportional to the number of electrons in an atom—that is, to the atomic number. For this rule would be true only if atoms were small, but they are not small compared with the wavelengths of X-rays nor—what is more to the point—with the spacing in crystals in general.

From fig. 4.9 we can see that the way in which the waves from two different points in an atom combine depends both on the separation of the two points and upon the angle of scattering. If we know the distribution of electrons in an atom we can see how the scattering varies with angle. The calculations are difficult but the general pattern of the results is easy to see.

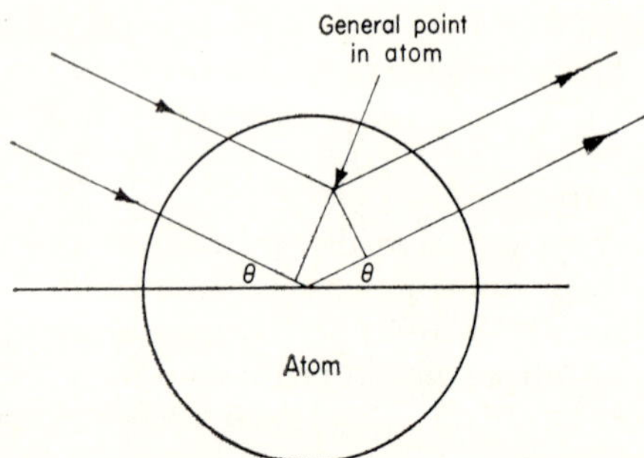

FIG. 4.9. Path difference for wave scattered by general point in atom relative to that scattered from centre.

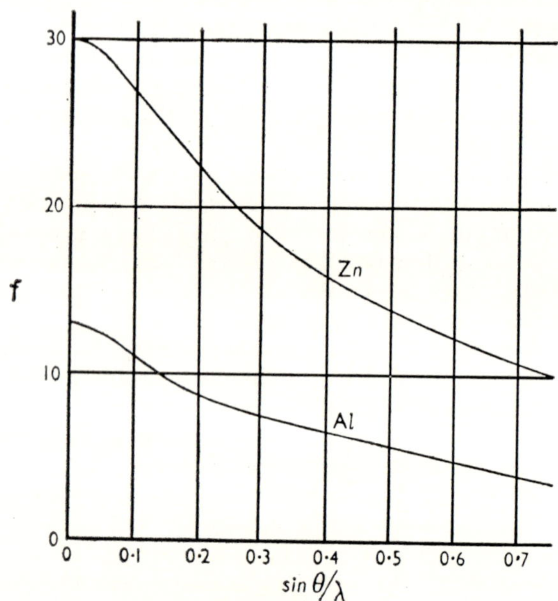

Fig. 4.10. Scattering factors of two atoms, showing decrease with increase of Bragg angle.

1. At small angles, the scattering is proportional to the number of electrons—that is, to the atomic number.
2. At larger angles, the scattering decreases (fig. 4.10) tending to zero asymptotically.

3. The scattering falls off more slowly for heavier atoms because these atoms have greater concentrations of electrons near the nucleus; in other words they behave as closer approximations to point atoms.

The curves representing the amplitude of the radiation scattered as a function of $\sin \theta/\lambda$ are known as *atomic scattering-factor curves*; the curves for all the different atoms are a necessary starting point for all work on X-ray diffraction.

Let us now build a crystal by putting an atom at each lattice point. Some elements do build structures in this way; they have one atom in the unit cell. We can work out the value of $\sin \theta/\lambda$ for any order of diffraction, and the amplitude of the radiation scattered will be proportional to the atomic scattering factor at that value.

4.8 *The effect of temperature*

If, however, we calculate the intensities of the various orders of diffraction from such a crystal and compare them with the observed values—after correcting for two factors that we shall consider in the next section—we find that there is a systematic discrepancy; the observed values are weaker than they should be, by a factor that increases as the angle of diffraction increases. The reason for this effect is that we have assumed that the atoms are stationary at the lattice points, whereas in fact they possess energy of heat motion and are vibrating rapidly. How do we allow for this motion?

It is tempting to regard the atom as being blurred out and thus occupying a larger volume than if it were stationary, and in fact this assumption gives the right answer. Nevertheless, the oscillations of the atoms, rapid as they are ($\sim 10^{12}$ Hz), are slow compared with those of X-rays (10^{18} Hz) and to a pulse of X-rays the atoms would simply appear as displaced from the exact lattice positions. The general effect of the two pictures—the average blurring and the X-ray 'snapshot'—works out the same, however.

A detailed treatment shows that the effect is not as simple as we have indicated. The atoms are not merely displaced at random from the lattice points; they are connected by interatomic forces which also govern the mechanical properties of the crystal—Young's modulus, rigidity and bulk modulus—and a most surprising result is that these elastic moduli can be derived from X-ray diffraction measurements. The subject is, however, too complicated to be discussed here.

4.9 *Correcting factors*

The two correcting factors that were mentioned in the last section are concerned with, first, the relative rates at which the crystal passes through its reflecting orientation, and, secondly, the fact that the diffracted radiation is partly polarized. The theory of the first factor was worked out by Lorentz, and leads to the result that reflections with small Bragg angles and large Bragg angles are enhanced with respect to the others; the *Lorentz factor* can be worked out and its values tabulated. The *polarization factor* is simple; if the incident radiation is unpolarized, the component parallel to the reflecting planes is fully reflected, but the amplitude of the perpendicular component is reduced by a factor cos θ. The intensity of this component is therefore proportional to $\cos^2 \theta$, and the average diffracted intensity is therefore proportional to $1+\cos^2 \theta$.

The Lorentz and polarization factors are usually combined together, and standard tables give values of the resultant factor as a function of θ. Each observed intensity must be divided by the corresponding value of this function.

4.10 *Diffraction by more complicated crystals*

We have now reached the stage of being able to cope with the theory of diffraction by a crystal with one atom at each lattice point. Very few crystals are like this; the vast majority have several atoms in the unit cell, and the simple theory that we have so far considered has to be modified. The waves scattered by the separate atoms in the unit cell interfere with each other in such a way that the amplitudes of the different orders of diffraction vary considerably; if the phases are such that the waves reinforce each other, the reflection will be strong, but otherwise it will be weak. Some reflections may be so weak that, although Bragg's law is obeyed, no reflection at all is observed.

The problem that we have to solve is to find how to add together the waves scattered from the atoms in the unit cell for all the conditions specified by Bragg's law. The fact that Bragg's law has to be obeyed reduces the problem to manageable proportions; otherwise we should have to find the scattering function of the atoms for *all* angles of diffraction for *all* orientations of the crystal—a gargantuan task indeed!

Even so, the problem is formidable, but theory has led to reasonably simple expressions that can be evaluated for many crystals without too much difficulty. Nowadays, in fact, digital computers can be used so that extremely complicated crystals can be dealt with. It is not our intention to derive these expressions here but simply to show the physical basis on which they rest. This basis can be explained most simply in two dimensions.

Suppose that our two-dimensional crystal contains four similar atoms in the unit cell; one is at the origin, O (and repeated at the other corners of the unit cell for convenience) and the others are at arbitrary positions, A, B and C (fig. 4.11 *a*). How can we find the resultants of the waves scattered by these four atoms? We consider each reflection separately. For example, take the planes ($\bar{1}20$) (fig. 4.11 *b*). (Because the third dimension is ignored, the third index (p. 34) is always zero). We know that, since the condition for production of a reflection is that all the corners scatter in phase, then all atoms lying on the planes passing through the corners also scatter in phase with each other. Therefore atom C scatters very nearly in phase, but A and B, which are almost exactly half-way between these planes, scatter almost exactly out of phase. Thus the scattering from O and C will be almost neutralized by that from A and B; thus the $\bar{1}20$ reflection will be very weak.

Take another set of planes 210 (fig. 4.11 *c*). Now we can say that B and C scatter almost in phase with O and that A is almost out of phase.

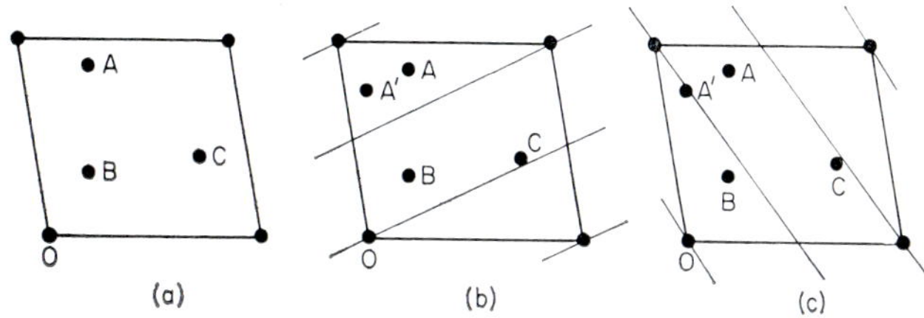

Fig. 4.11. (*a*) Positions of four atoms—O, A, BC—in two-dimensional unit cell; (*b*) positions of atoms relative to ($\bar{1}20$) planes; (*c*) positions of atoms relative to (210) planes.

The total effect is therefore equivalent to the scattering of two atoms—a moderately strong result. If A were moved to A′, it would make practically no difference for $\bar{1}20$, but would increase 210 almost to the maximum possible.

As we have said, there are general formulae that give these answers without geometrical construction. The formulae include the indices *h k l* of the reflections, and the coordinates, *x y z*, of the atoms; it is

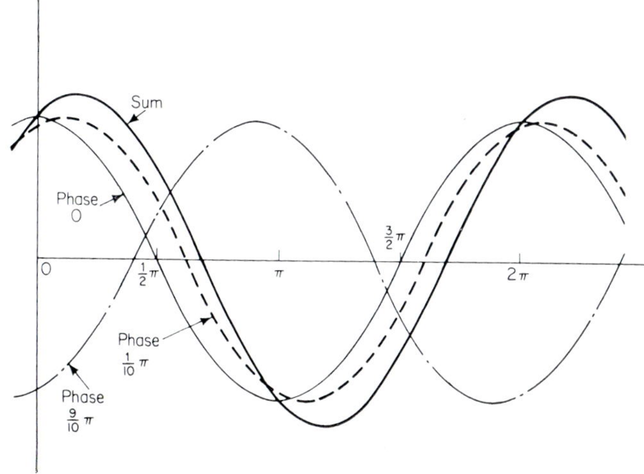

Fig. 4.12. Addition of three sinusoidal waves with different phase angles, showing that the sum is not greatly different from the contribution of one atom.

convenient to express these coordinates as fractions x y z of the cell edges, not as distances. For example, the coordinates of the atom B are roughly 0·25, 0·33, 0·00.

The resultant is expressed as a sine wave, whose amplitude is proportional to the amplitude calculated in the way that we have described. This is called the *structure amplitude*. But the amplitude does not specify completely a sine wave of known wavelength; we need to know the *phase* as well. Figure 4.12 shows three waves of the same amplitude that differ in phase. The second is slightly in front of the first, and the third is nearly half a wavelength in front. We say that, if the first wave is the standard, the second has a small positive *phase angle* and the third a large one. If a complete wave corresponds to 2π radians, the second wave has a *phase angle* of about $\pi/10$ and the third a phase angle of $9\pi/10$ (or $-11\pi/10$ if one wants to look at it that way).

The derivation of the structure amplitude that we have just described is really the addition of waves of this sort (compare § 1.6), one for each atom. Some will be nearly in phase and some nearly out of phase, as we have shown. The resultant will be a wave with a phase angle α related to that scattered by the atom at O. Even if there were no atoms at O, it would still be convenient to measure the phase with respect to the scattering that would have taken place by an atom at O.

The complete wave is thus specified by an amplitude and a phase angle. This combined quantity is called the *structure factor*. It is represented by the symbol $F(h\ k\ l)$, since it has a separate value for each reflection.

4.11 *Complete representation of a diffraction pattern*

We have now dealt with all the processes involved in the diffraction pattern of a perfect crystal. The reciprocal lattice is a collection of points representing all the possible reflecting conditions, and if we attach to each point two symbols, representing the structure amplitude and the phase angle of the appropriate reflection, we have a complete representation of the diffraction pattern. The aim of the rest of this book is to explain how far we can go towards determining this representation experimentally, and what use we can make of the information when we have it.

CHAPTER 5
experimental arrangements

5.1 *General view of problems*

For investigating the structure of a crystal, the basic requirement is simple enough: we merely wish to record all the orders of diffraction that can occur. Objects with one-dimensional and two-dimensional periodicity would present no difficulty; we allow a beam of radiation to fall on them and the diffraction patterns can be recorded on a photographic plate. For the reasons outlined in the last chapter, however, the same procedure is not applicable to crystals; we cannot record a three-dimensional diffraction pattern on a two-dimensional film. The present chapter will be concerned with describing the various procedures that have been used, showing how a gradual increase in complexity and apparatus has developed in order to simplify the derivation of the diffraction pattern.

5.2 *Laue method*

As we showed in Chapter 2, the first X-ray diffraction photograph was taken with the simplest possible experimental arrangement—a fine X-ray beam falling upon a stationary crystal. This method was much used in early work and it now seems amazing how much ingenuity was applied to analysing the resultant photographs, which were called *Laue photographs*.

To produce a fine beam of X-rays, we need a long hole in a cylinder made of a heavy metal such as lead; this metal is particularly useful because it can easily be cast round a straight rod. There is, however, one difficulty with this arrangement; the end of the hole can produce unwanted diffraction effects (fig. 5.1 *a*) and to eliminate these as far as possible it is usual to widen the hole at the end (fig. 5.1 *b*) so that most of this diffracted radiation is prevented from reaching the film.

The crystal can be supported in any orientation. But if symmetrical

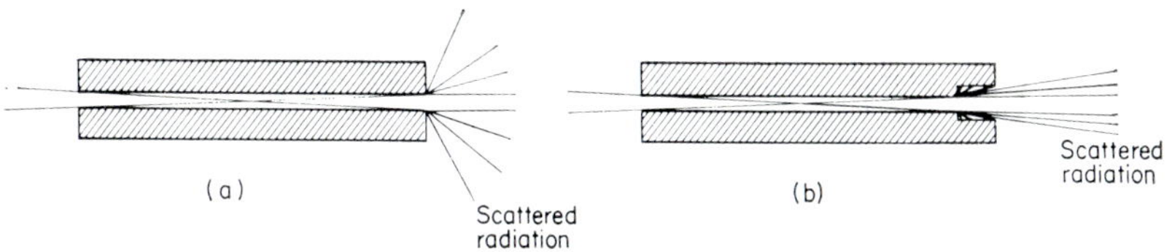

Fig. 5.1. (*a*) Simple collimator; (*b*) collimator with recess to reduce scattering from end.

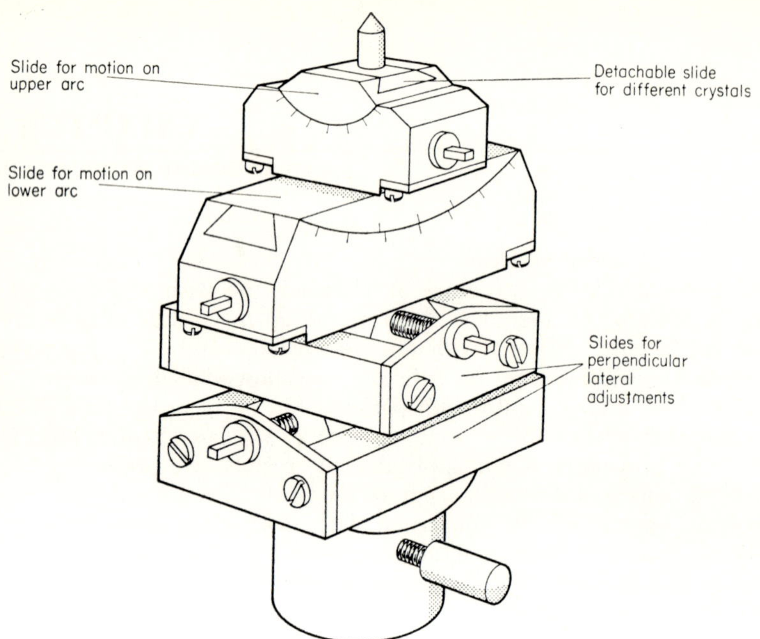

Fig. 5.2. Set of arcs for supporting crystal so that it can be translated and orientated in two dimensions. (Courtesy of W. Hughes.)

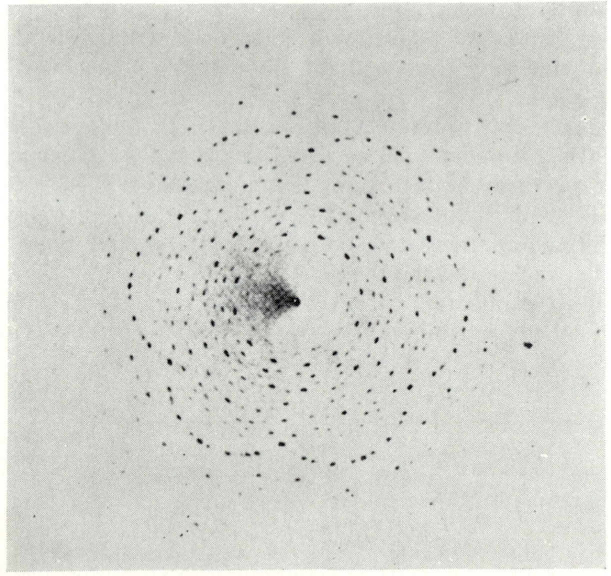

Fig. 5.3. Laue photograph from stationary crystal of beryl.

photographs (fig. 2.5) are wanted it is necessary to be able to adjust the orientation of the crystal by measured amounts and it is therefore usual to support the crystal on goniometer arcs (fig. 5.2); these allow adjustment through two known angles in perpendicular planes. This type of support is important because it is common to all types of single-crystal apparatus.

With a properly adjusted crystal, the spots—recorded on a flat film—lie on curves arranged in symmetrical ways (fig. 5.3); there is no doubt that Laue photographs of crystals of high symmetry are the most beautiful types of X-ray diffraction photographs. But unfortunately beauty and utility do not necessarily go together, and nowadays Laue photographs are used only for preliminary examination of symmetry in special circumstances—for example, when a crystal forms part of a large conglomerate.

The disadvantages of Laue photographs have already been mentioned (p. 52). The spots are formed from different wavelengths of the continuous spectrum (p. 22) and so their intensities are not easily comparable; a spot produced by a wavelength which happens to be near the peak of the intensity distribution will be stronger than one with the same structure amplitude (§ 4.10) produced by a wavelength well away from the peak. Therefore a knowledge of the intensity as a function of wavelength is needed.

Secondly, all the orders of diffraction from the same set of planes (p. 32) overlap, as we can see from Bragg's equation, $n\lambda = 2d \sin \theta$. Since the rays are incident at a given angle θ on to the planes, the law is obeyed for $n = 1$ and a particular wavelength λ; it is therefore also obeyed for $n = 2$ and wavelength $\lambda/2$, for $n = 3$ and wavelength $\lambda/3$, and so on down to the smallest wavelength existing in the X-ray beam. Thus a given spot on a Laue photograph will have contributions from all these orders with different wavelengths (harmonics).

The early workers in the field devoted much attention to overcoming these defects. By reducing the voltage on the X-ray tube, the short-wavelength limit (§ 2.8) could be raised, and so the high orders, reflecting on the shorter wavelengths, could be removed one by one. In this way it was possible to determine the relative intensities of the separate orders of diffraction. But such methods were not very reliable; there were too many possibilities for errors and inaccuracies.

One particular problem was caused by the presence of characteristic radiation (p. 22 and fig. 2.8); if a set of planes happened to reflect this radiation, the resultant would be exceptionally strong and its intensity could not be compared with those of other spots. This disadvantage, however, gave rise to another possibility, which is now used universally in serious studies of crystal structure.

5.3 *Rotation and oscillation methods*

If the crystal is rotated, as we saw on p. 50, the orders of diffraction of a fixed wavelength flash out as the crystal planes pass through their reflecting orientations. Why not then use these orders of diffraction which are produced by the same wavelength, and for which the different reflections from the same planes occur quite separately? This suggestion was made by Schiebold in 1919, and has been one of the most fruitful ideas ever injected into the practical side of the subject.

We can see the relationship between the Laue photograph and the rotation photograph by considering what happens if we change slightly the orientation of a stationary crystal: the spots on the Laue photograph, since they represent reflections from planes, move slightly. Thus, if there is continuous movement, the Laue spot will trace out a streak as it moves across the film. The streak represents a variation of wavelength, and when the wavelength happens to be a characteristic radiation, there is a sudden enhancement of the streak. These enhancements are the characteristic spots on which all modern work depends.

There is a slight disadvantage: there is not just one characteristic radiation from an X-ray tube; there is a sequence $K\beta$, $K\alpha_1$ and $K\alpha_2$ in increasing order of wavelength. The β radiation can be eliminated by passing through a material that absorbs the β more than the α's. For every target element there is usually another element that has this property; for example, nickel absorbs $CuK\beta$ much more than $CuK\alpha$, and a thin foil, about 0·02 mm thick, in these incident beams will eliminate the $K\beta$ radiation completely.

The α radiations present a slight difficulty; these wavelengths are very close, differing only by about one part in 400 for a target such as copper. At small angles, the α spots record as one and there is no problem, but at high angles they produce a close doublet, which can be seen on many of our illustrations. Even with quite crude apparatus the two radiations can be resolved if θ is near 90°, and the effect introduces some complexity into the measurement of intensities at these angles.

Otherwise, the method works well, particularly if the crystal is made to rotate around the direction of one of the axes of the unit cell. Then one great simplification occurs; if the crystal is surrounded by a cylindrical film whose axis is the axis of rotation (fig. 5.4), the spots lie upon straight lines when the film is laid flat. These lines are called *layer lines* and play an extremely important part in the analyses of the photographs.

For, if the axis is along the direction of a cell edge, we can regard the crystal—whatever the symmetry—as a set of planes repeated at regular intervals along the axis (fig. 5.5 *a*). Let us suppose that this interval is c, one of the unit-cell parameters (§ 3.4). One condition for reinforcement (p. 44) is that $c \sin \phi_n = n\lambda$ (fig. 5.5 *a*). Thus, whatever the intervals along the other two axes are, the angles ϕ_n must assume a set of values corresponding to l (§ 4.4) $= 0, \pm 1, \pm 2, \ldots$ These values must intersect the cylinder in a set of circles perpendicular to the axis, and when the film is laid flat these become straight lines (fig. 5.5 *b*). In fact, these lines represent the solution of one Laue equation (p. 44).

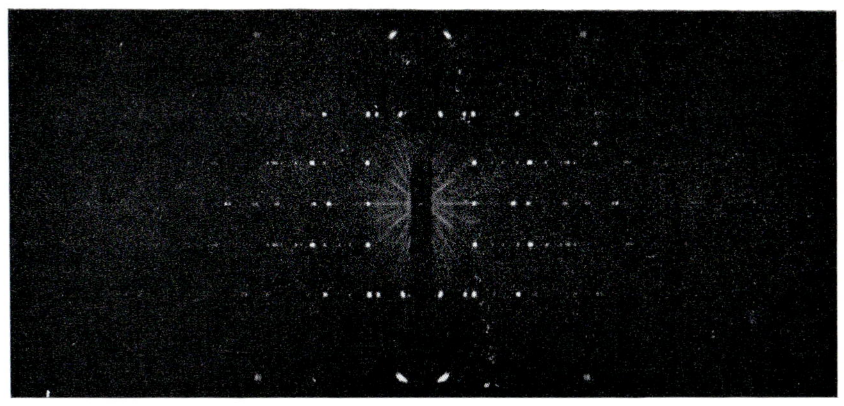

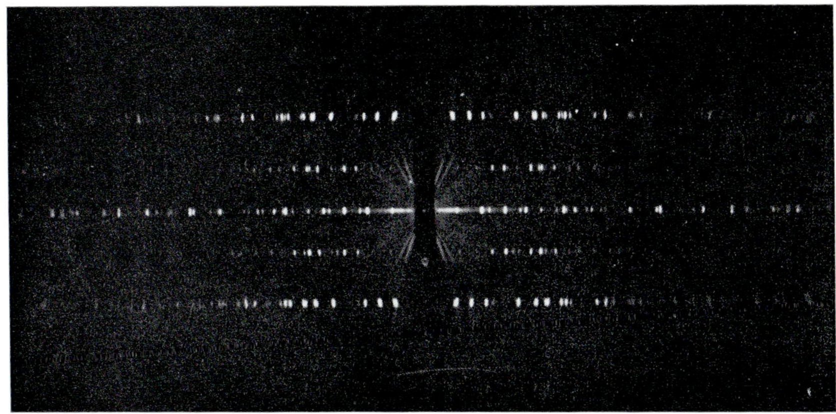

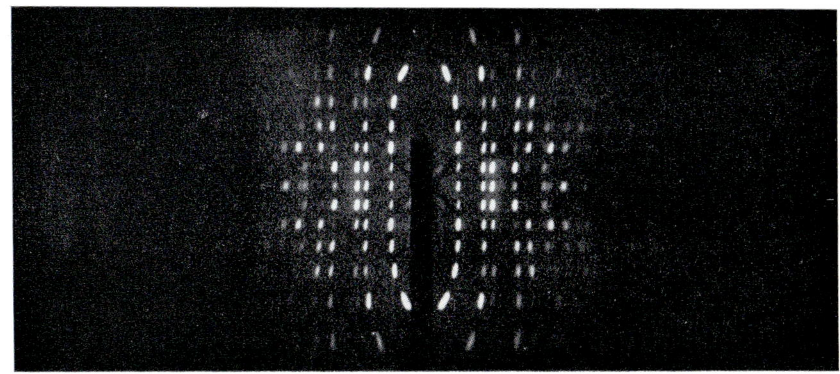

Fig. 5.4. Rotation photographs showing straight layer lines.

The first result of this fact is that we can derive the value of the c edge of the unit cell directly, with a reasonable degree of accuracy—about $\frac{1}{2}$%. We also know that all the spots on the zero layer line—the central one—have $l = 0$, those on the first layer line have $l = 1$, and so on. This however is all the information that we can derive directly; there is no simple way of finding the other two cell dimensions, or of finding the other two indices of the spots.

The difficulty is that we deprive ourselves of one essential piece of information when we rotate the crystal; we have no idea of the orientation of the crystal when a particular reflection is being produced. This difficulty can be reduced by the obvious expedient of oscillating the crystal through a small angle—say 10°—instead of rotating it. The principle remains the same, but now we know the orientation of the crystal within 10° for each spot. This is of great help. It can still be difficult to find the unit cell, but if it is known, fairly straightforward methods can be used to find the indices of the spots. A paper by J. D.

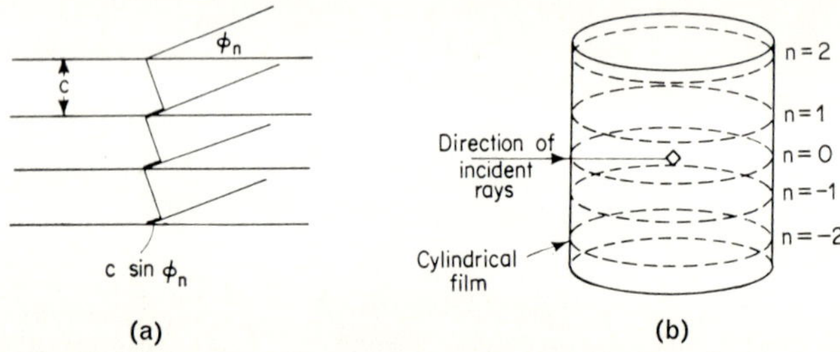

Fig. 5.5. (*a*) Crystal represented as set of planes, with one condition for reinforcement; (*b*) production of layer lines, $n = 0$, $n = \pm 1$, $n = \pm 2$.

Bernal, working with W. H. Bragg in London, in 1926 played an important role in systematizing the procedure, and from that time on his method was used by everyone engaged in interpreting what are called *oscillation photographs*.

This method made use of a geometrical concept introduced by Ewald in 1913. It can be simply explained in terms of diffraction by a one-dimensional grating, the reciprocal lattice of which, as we explained on p. 51, is a set of parallel lines separated by a distance λ/d (fig. 5.6). For the Ewald construction we select as origin a point O on the central line of the reciprocal lattice. We then draw a circle of radius 1 unit, matching the scale of λ/d; the circumference of this circle must pass through O, and the diameter through O must be in the direction of the incident beam. (Note that λ/d, being the ratio of two lengths, is dimensionless, and therefore all distances in the reciprocal lattice are also dimensionless.) From fig. 5.6 it can be seen that the directions of the diffracted beams can be obtained by joining the centre of the circle to the points where the circle cuts the reciprocal-lattice lines.

It would, of course, be absurd to use this construction, simple though it is, to solve the problem of diffraction by a one-dimensional grating illuminated normally. Its usefulness lies in the fact that it can be applied to non-normal incidence and also to two- and three-dimensional diffraction; it shows clearly the properties of the process of diffraction described in § 4.4.

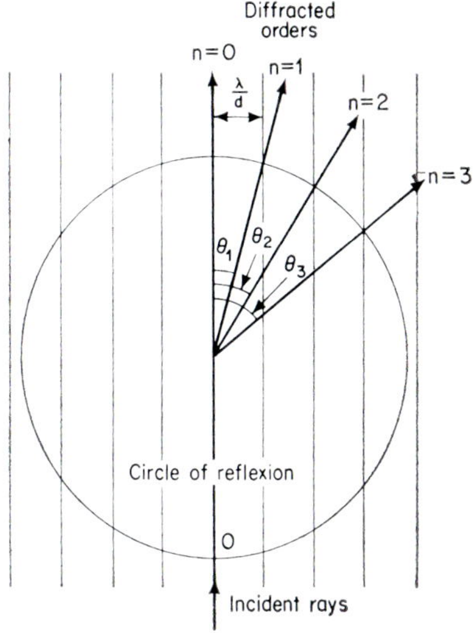

Fig. 5.6. Reciprocal lattice of one-dimensional diffraction grating, with circle of reflection.

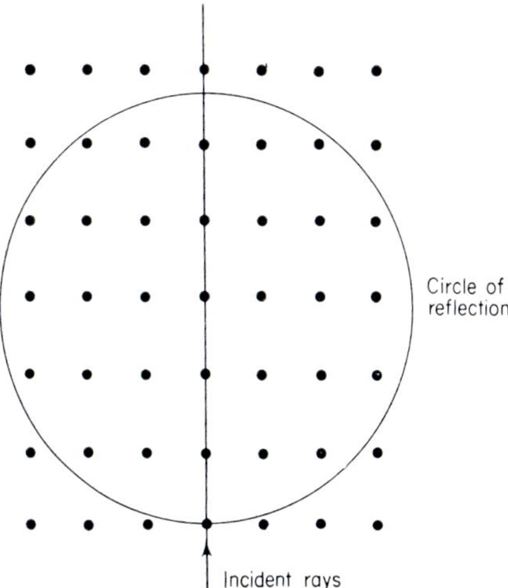

Fig. 5.7. Reciprocal lattice of two-dimensional grating, showing circle of reflection not passing through any reciprocal-lattice points.

In three dimensions, the circle becomes a sphere, but we can illustrate its operation with a two-dimensional figure (fig. 5.7). We represent the reciprocal lattice by a collection of points and the sphere by a circle; in any random orientation of the sphere, the circle is unlikely to pass through any of the points, corresponding to the fact that, with monochromatic radiation, no orders of diffraction, except of course the zero order, will occur (p. 49). If, however, the reciprocal lattice is rotated, corresponding to crystal rotation, a reflection will flash out each time a reciprocal-lattice point passes through the surface of the sphere.

To identify which reflections can occur with a given range of oscillation of the crystal, one merely draws two circles in the reciprocal lattice

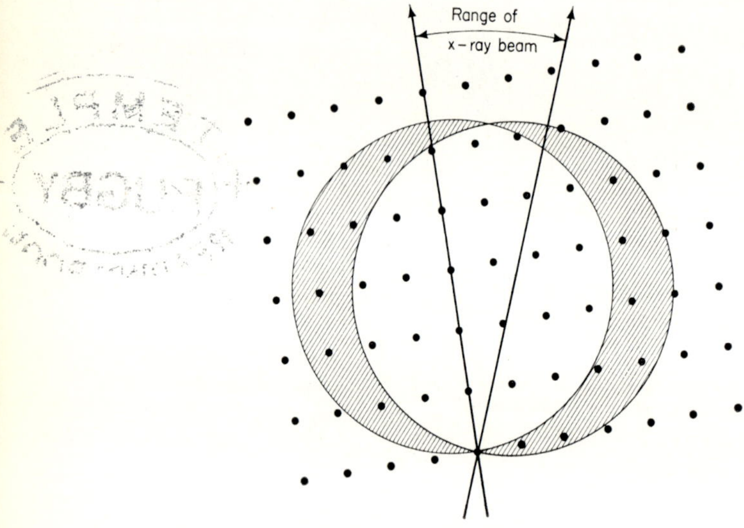

Fig. 5.8. Use of two circles of reflection to represent extremes of oscillation. All reciprocal-lattice points in the shaded area represent possible reflections.

representing the extreme ends of the range (fig. 5.8); any point lying within the area traced out by the circles represents a possible order of diffraction. Bernal described a systematization of this procedure and its extension to three dimensions.

Bernal called the sphere 'the *sphere of reflexion*', but it is often called 'the *Ewald sphere*' in honour of its inventor. As the reciprocal lattice rotates in all possible orientations, the Ewald sphere traces out a sphere of radius 2 units and this is called the *limiting sphere*, representing the fact that, from Bragg's law (equation (4.1)), the maximum value of λ/d is 2.

The oscillation method has its drawbacks. First, to cover complete rotation—or even 180°, which would do just as well—a large number of photographs has to be taken. More than eighteen 10-degree photographs are needed because there has to be some overlap so that some identical spots appear in successive photographs; these serve to correlate the intensities, because one cannot guarantee to keep conditions of exposure and development constant from one exposure to the next. Moreover, Bernal's method was not always quite unambiguous, and the procedure was lengthy so that mistakes could arise. Other methods were therefore sought.

5.4 *The Weissenberg method*

A method had been suggested, in 1924, by the German physicist, Weissenberg, but it had not been generally adopted because, with the relatively simple crystals then being investigated, the oscillation method was not too onerous; people were reluctant to replace an acceptable

WITHDRAWN
TEMPLE
RUGBY
READING ROOM

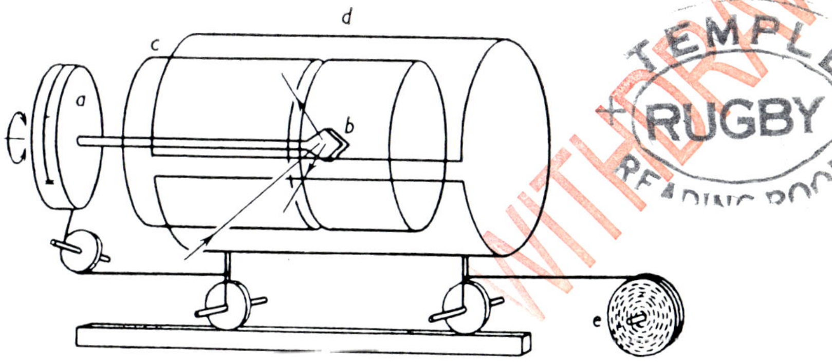

Fig. 5.9. Mechanism of Weissenberg goniometer. *a* represents the arcs, *b* the crystal, *c* the screen and *d* the film

procedure by one that required the building of new and rather complicated apparatus. Fortunately, there are always some scientists who are interested in instruments rather than in results, and so a few Weissenberg goniometers were built. As soon as their advantages became obvious, they were adopted in laboratories all over the world; they are now the instruments of most general utility in crystal-structure determination.

The basic idea is very simple: since we cannot record a three-dimensional diffraction pattern on two-dimensional film, let us reduce the problem to two dimensions by exposing one layer line at a time. We know the l index of the spots on each layer line, so that there are only two indices to find. The layer line is allowed to pass through a narrow space in a screen, and all the rest of the pattern is eliminated. The film is then drawn past the layer-line screen during the exposure, so that the spots are recorded on different parts of the film (fig. 5.10).

The motion of the film has to be synchronized with the rotation of the crystal. It is usual to oscillate the crystal through 180°—or rather more than 180° to avoid discontinuities at the ends of the oscillation—and the film has to move accurately up and down through a given distance for each complete oscillation of the crystal.

The theory is simple; the distance along the film perpendicular to the layer-line screen gives us a measure of the angular position of the crystal when the particular spot was being produced. This information, together with the Bragg angle derived from the measurement parallel to the layer-line screen, is sufficient to identify each reflection unambiguously.

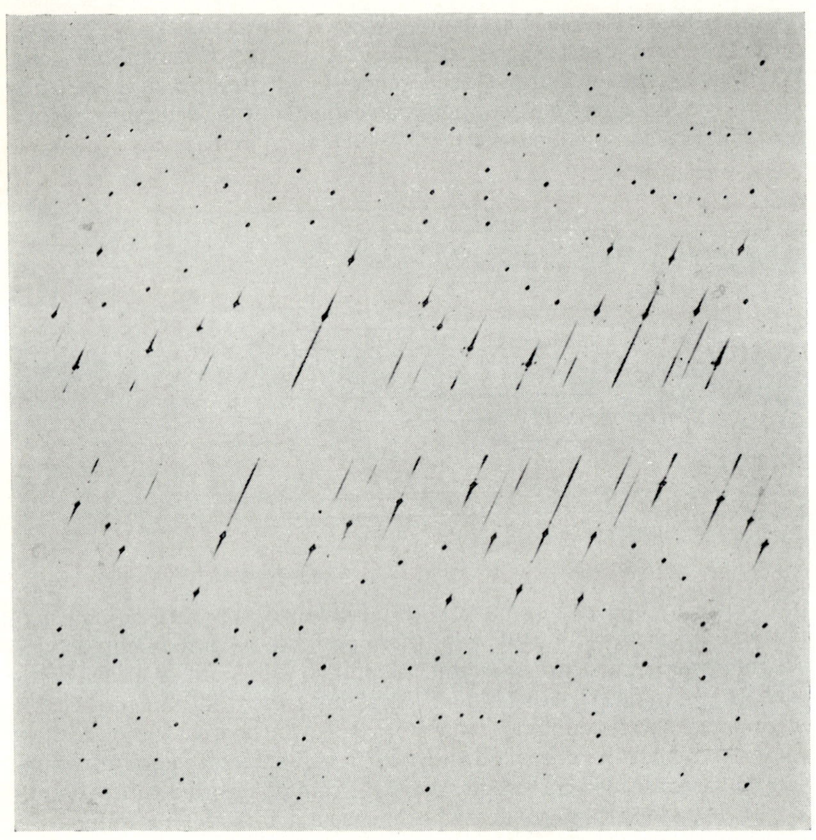

Fig. 5.10. Weissenberg photograph.

In fact, the operation turns out to be much simpler than this. Weissenberg photographs, of which fig. 5.10 is an example, have patterns of spots strung out along beautiful curves. These curves turn out to be lines in the reciprocal lattice (remember that the third index is already fixed). Certain lines (fig. 5.10) are straight: these represent the lines in the reciprocal lattice for which one of the unknown indices is zero. These represent the axes of the reciprocal lattice, and it is then not difficult to read off the indices of all the other spots. It takes very little training to index a Weissenberg photograph merely by inspection, but to carry out the operation properly, particularly for crystals with large unit cells, charts are available which enable the Weissenberg lines to be drawn for any unit cell.

By taking photographs for the different layer lines, most of the diffraction pattern can be recorded; photographs with the crystal oscillating about the other two axes can supplement and confirm the information so derived. In these ways the whole three-dimensional diffraction pattern can be built up.

5.5 *The precession method*

It might seem, then, that the Weissenberg method is the ultimate answer for deriving diffraction patterns of crystals; it is simple and unambiguous, and it would appear to satisfy all the requirements for providing the data for working out crystal structures. There is, however, a slight disadvantage: for crystals with large units cells the spots are very close together, and the allocation of indices is sometimes rather dubious, particularly for high orders. For this reason several attempts were made to record the spots exactly in their reciprocal-lattice arrangement, when there could be no doubt at all. The most successful of these devices is the *precession method* described by Buerger, of Massachusetts, in 1939.

In this method the crystal undergoes an odd motion that we shall not attempt to describe, and the diffraction pattern is recorded on a flat plate undergoing a similar odd motion. One layer line at a time is recorded; screens with different sizes of annular circular openings are provided with the instrument to extract the layer line required. The instrument is fascinating to watch; the several parts move with different relative motions, and they seem bound to foul each other. But they just manage to miss!

Figure 5.11 shows the precession photograph of a protein crystal, which has a unit cell much larger than it is possible to deal with by the oscillation and Weissenberg methods. It might therefore be asked why the Weissenberg method is still used at all; why is it not completely superseded by the precession method? The answer is that, because the method requires the use of a flat plate, it cannot, in principle, record reflections with Bragg angles greater than 45°; in practice the limit is about 30°. This limit would be unacceptable for most work.

The difficulty can, however, be overcome by the use of X-rays of a shorter wavelength; the Kα radiation of molybdenum, with a wavelength less than half that of CuKα, is very popular. The solution however is not ideal. The pattern is compressed so that the spots are very close together, and this may make the assessment of the intensities (§ 5.6) rather difficult. Moreover, X-ray film is less sensitive to radiation of shorter wavelengths because the absorption is less; longer exposures are required for MoKα than for CuKα. For these reasons, the precession method is not normally used unless the research requires it; the Weissenberg method, which can record spots with Bragg angles up to nearly 90°, still holds the field.

5.6 *Measurement of intensities*

Allotting indices to the spots in our photographs is only the first step in our problem; we also have to measure the intensities. There are several ways of making these measurements, but most of the early work on crystal structures was carried out by visual estimation. It must seem surprising to students who have been brought up to believe that 'science is measurement' to accept that serious scientific results could be obtained from measurements made so roughly, but, of course, a reading of the

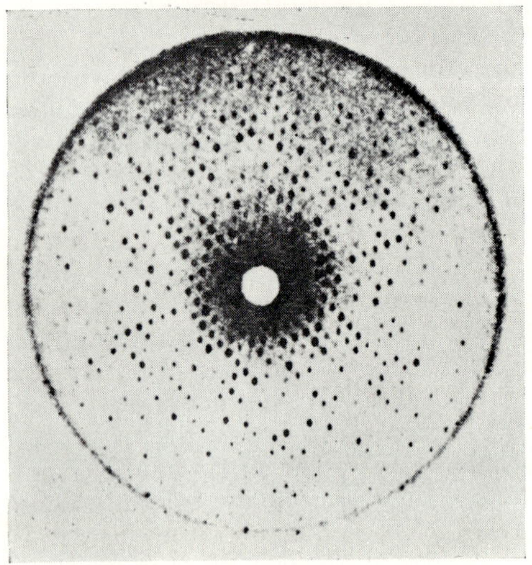

Fig. 5.11. Precession photograph.

history of science shows that practically every subject started in this way; exact measurement was used only when the ground-work had been laid by rough methods.

Visual estimation merely amounts to classifying spots as strong, medium and so forth. A standard strong spot would be chosen as one which was quite black, but not so strong that some diffuseness surrounded it; a medium one was of average intensity, quite clear, but not black; a weak one was one that was below average, but yet clear enough to be seen without close examination. These classes could be supplemented by extra divisions—very strong, medium-strong, weak-medium, very weak (observable only with difficulty), and of course, zero. The total of eight classes—v.s, s, m-s, m, w-m, w, v and o—was sufficient to enable structures not only to be determined but to be determined with considerable accuracy. (In German papers the symbol s meant schwach—weak—not strong, and this could be a source of considerable confusion to English-speaking crystallographers!)

The reason for the accuracy of this apparently rough work is not difficult to see. Suppose that we are studying a crystal with a cell edge of about 8Å (the Å is defined in § 6.3), which will give about 10 orders with Cu Kα radiation ($\lambda = 1{\cdot}54$ Å). A shift of one-hundredth of the unit cell will produce a phase change of $10 \times 2\pi/100$ in the 10th order. This is 36°. The angle is sufficient to change a cosine from 1 to 0·8, or from 0 to 0·6. These changes are large, and will affect the structure amplitude (p. 58) considerably—well above the limits set by our rough estimations. We can therefore fix our atoms to better than $0{\cdot}01 \times 8$ Å—say 0·05 Å. Recent refined work on some early structures shows that they were indeed more accurate than this.

Gradually, however, more quantitative methods were introduced. Scales of spots of known intensity ratios were used; they were made by allowing a strong reflection from a crystal to be recorded by specified numbers of passes through its reflection orientation, a row of spots being formed on a separate piece of film. This could then be used for quantitative comparison with the observed spots. Some people acquired a considerable degree of skill in this activity and could work to an accuracy of better than 20%; in fact, later precise measurements sometimes showed that the accuracy was even better than this. It should be noted that an error of 10% in intensity is equivalent to an error of only 5% in structure amplitude, since errors are halved when square roots are taken (p. 8). Since there are some errors, which we shall not discuss here, which could be much greater than 5%, it seemed to be hardly worth while trying to improve the accuracy further.

Nevertheless, by the aid of the computer, these errors could be corrected and then precise measurements become worth while. In addition, it was unsatisfactory to have to depend upon subjective measurements, which not everybody could carry out with the same degree of skill, and thus some simple instruments began to be built. The simplest type involved the passage of a narrow beam of light through the spots, the intensities being measured by the emission from a photoelectric cell. These instruments, however, introduced many sorts of complications. The total intensity was needed, not just the peak. Therefore one had to have finely spaced readings over each spot, and the total readings had to be added together. But this required knowledge of the relationship between the initial incident intensity of the X-ray and the absorption properties of the film. The whole subject became complicated, and the measurements became impossible without the use of automatic instruments. This is the way of the world, unfortunately; when we depart from pristine simplicity, difficulties—seen and unseen—multiply. It is not certain that the problem has been completely solved yet.

5.7 *Automatic diffractometers*

The most accurate methods now dispense with film entirely. Film is used for preliminary work—determination of unit cells, detection of symmetry and so forth—but for estimation of intensities we now revert to what is, essentially, the ionization spectrometer (fig. 2.7). Now, however, we do not measure the current produced by the ionization; we count the number of photons by means of some device such as a Geiger counter. The rate of arrival of photons is a direct measure of the intensity of the X-ray beam.

It is possible to make the counter record its observations automatically. This facility is however of little use unless the crystal can also be set to pass automatically through all its reflecting orientations. Much ingenuity has been spent on devising instruments that carry out this operation; these are called *four-circle diffractometers*. Only two of these circles correspond to the axes on which the crystal is supported (p. 60) and the third corresponds to rotation about the axis. The fourth circle is needed in order to adjust the position of the counter automatically at twice the angular velocity of the crystal, in order to keep the Bragg condition for reflection satisfied.

The procedure, then, in examining an unknown crystal is first to take single-crystal photographs—Weissenberg or precession—to establish the unit cell and space group (p. 38). With this information the crystal is mounted and the diffractometer is instructed to set itself to all possible reflecting orientations and to measure the intensities of the resulting orders of diffraction. The results are all recorded.

When the instrument works well, it solves all the problems of obtaining diffraction patterns of crystals. It will, however, readily be appreciated that such a complicated instrument will have many possibilities for faults and these are not readily rectified by an ordinary research worker. Thus a heavy price has to be paid for the considerable facilities that the diffractometers offer; the research worker is no longer in complete control of his work. For complicated problems this is a price that must be paid, but for more ordinary ones many people still prefer the simple instruments which they can manage themselves.

5.8 *Summary of single-crystal methods*

We have come a long way from the beginning of the subject. The simple Laue method enabled X-ray photographs to be obtained with quite crude equipment, but these photographs were difficult to interpret, and to obtain quantitative measurements from them was almost impossible. The use of characteristic radiation, reflected from a rotating or oscillating crystal, gave patterns that could be more readily interpreted, but the apparatus was, of course, rather more complicated. The task of interpretation was still, however, rather onerous, a large number of photographs was necessary to cover the whole of the diffraction pattern of a crystal.

The next break-through came in the form of what are called moving-film devices, in which the film moves as well as the crystal. With these, interpretation became very simple, but the apparatus was more complicated. The most popular device is the Weissenberg camera, and this is used for most crystal-structure work. For crystals with large unit cells the complexity of the Weissenberg diffraction pattern can lead to uncertainty, and then Buerger's precession method, which gave a direct representation of the reciprocal lattice, is more useful.

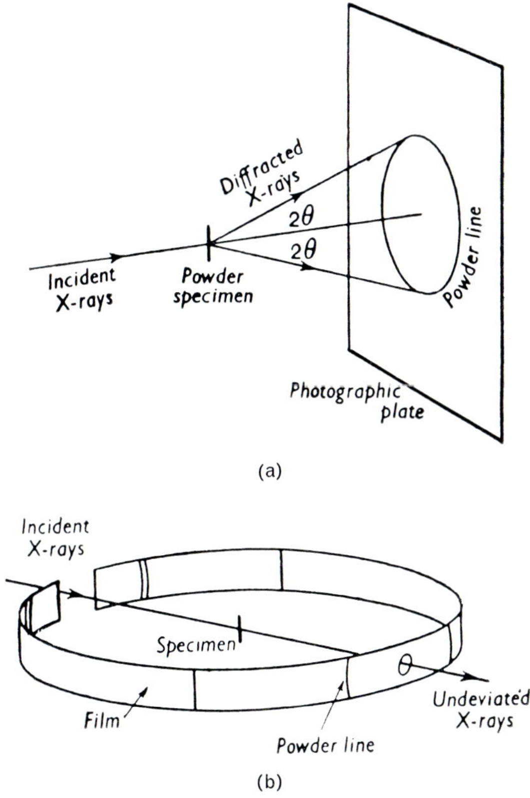

Fig. 5.12. (*a*) Production of reflections from powder, showing that they lie on cones of semi-angle 2θ; (*b*) recording of powder reflections on a strip of film.

This is probably the end of the story. This may seem a bold statement; at every stage of development of a subject, the present state seems, to most people, to provide all that is necessary. But in X-ray crystallography it is difficult to see what improvements can be envisaged in photographic methods, and, now that automatic diffractometers are available, it seems that photographic methods may ultimately be relegated to minor roles in the work. It is unlikely that they will be displaced altogether; there is no substitute for the overall view of the problem that one gains by looking at a Weissenberg or precession photograph.

5.9 *Powder photographs*

So far we have assumed that the material that we wish to study can be obtained in the form of a perfect single crystal. What can we do if it cannot? Some materials—for example metals and alloys—can normally be obtained only as conglomerates of very small crystals packed tightly together in random formation. Others exist only as fine powders. Are these outside the range of X-ray diffraction methods?

The answer is, 'Not necessarily'. At first sight it may seem that a random collection of small crystals would give an undecipherable diffraction pattern, but application of Bragg's equation (4.2) shows that this is not so; with monochromatic radiation each set of planes produces orders of diffraction with definite Bragg angles θ, and therefore deviations of 2θ. Thus, whatever the orientation of the planes, a reflection with specific indices must lie in a cone with a semi-angle 2θ with the incident direction as axis (fig. 5.12 *a*). If a cylindrical film is placed round the specimen, a series of lines will be recorded upon it (fig. 5.12 *b*).

Each line represents an order of diffraction. It may seem incongruous that we should have to conclude this chapter, which has been concerned with the problems of deriving a three-dimensional diffraction pattern from two-dimensional films, by dealing with a method that compresses the information into only one dimension, but that is simply what the *powder method*—as it is called—does.

The apparatus is now simple again. We need only a narrow strip of film to record the pattern, and the specimen can be stationary. In fact it is usually rotated, to bring as many crystals as possible into reflection orientation; a powder sample is not usually random enough to give smooth lines if it is stationary. (An automatic diffractometer—simpler than a single-crystal instrument—can also be used.)

Some typical powder photographs are shown in fig. 5.13. All of them are produced by relatively simple structures; obviously if the unit cell is large, the number of lines on a powder photograph would also be large, and it would not be possible to separate all of them from their neighbours. If a crystal is cubic, however, this result does not apply. It can be shown that the spacing of the (*hkl*) planes in a cubic crystal is given by the equation:

$$a_{h,k,l} = \frac{a}{\sqrt{(h^2+k^2+l^2)}}, \tag{5.1}$$

where a is the edge of the unit cell. Thus it follows that

$$\sin \theta_{h,k,l} = (\lambda/2a)\sqrt{(h^2+k^2+l^2)}. \tag{5.2}$$

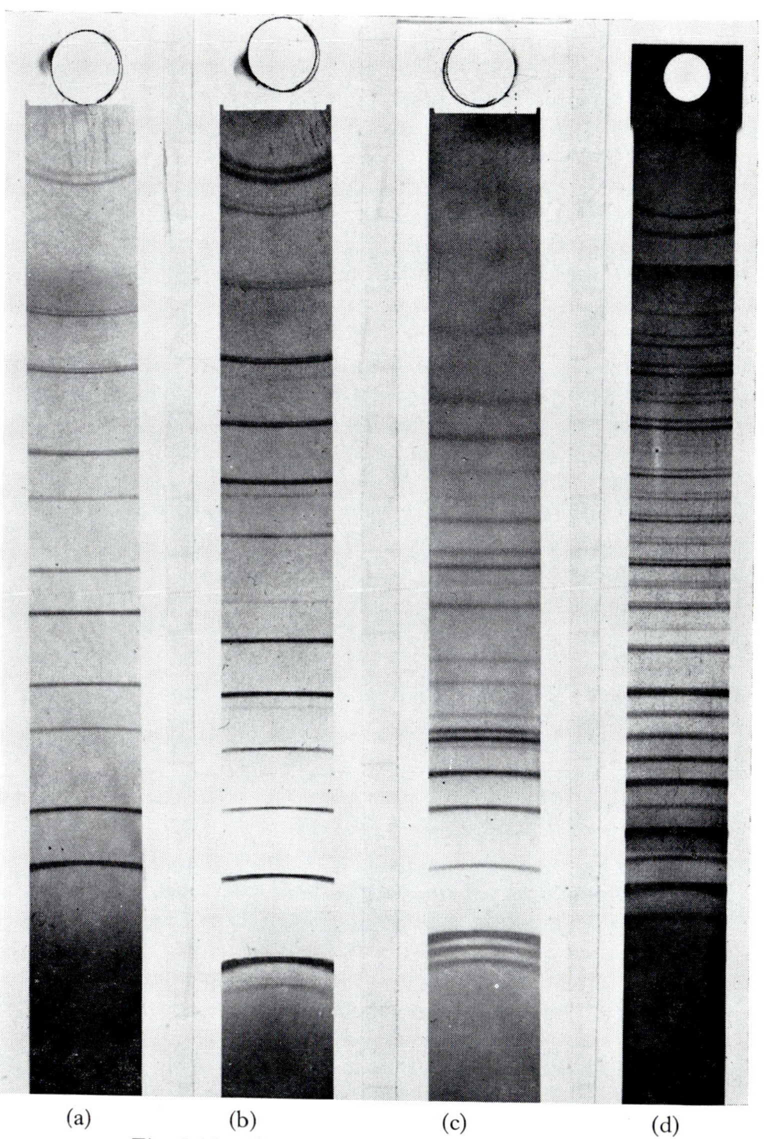

Fig. 5.13. Some typical powder photographs.
(*a*) silicon, (*b*) NaCl, (*c*) Zn (CuKα radiation), (*d*) Fe (MoKα radiation)

The quantity $h^2+k^2+l^2$ must, of course, be integral, and therefore the value of $\sin\theta_{h,k,l}$ can only have values simply related to $\lambda/2a$; thus the lines on the powder photograph of a cubic substance have a regular sequence (fig. 5.13), corresponding with the integral values of $h^2+k^2+l^2$. (It may amuse the reader to find which integral values cannot occur; for example, the sum of three squares cannot add up to 7. There is a simple formula relating these values.)

Powder photographs of non-cubic materials are more complicated, and it is not always possible to allot indices to the lines solely from the values of the observed Bragg angles. It may be wondered, therefore, what purpose these photographs serve. In fact, it turns out that, for purposes other than crystal-structure determination, powder photographs are more useful than any of the other types of photographs described.

First, because orientation is no longer involved, the powder photograph of a substance is characteristic of that substance and can be used to identify it. We do not need to know anything about the structure to use this method of identification, any more than we need to know a person's characteristics in order to identify him by his fingerprints. A great deal of work has been put into developing systems for this sort of work, in industry it is one of the most important applications of X-ray diffraction (see chapter 12).

Secondly, because the powder camera is so simple it can be made very accurate; by its means, lattice spacings can be measured to about 30 parts per million. This accuracy is attainable because of the large dispersion that occurs near 90°; the α doublet represents a variation of wavelength of only about 0·2% and yet it can lead to a separation of about 1 mm on a film (fig. 5.13). The positions of the lines can be measured to within 0·02 mm, leading to the accuracy stated.

This property is extremely useful in dealing with alloys. When two metals are melted together they often form intimate mixtures which, on solidifying, form a homogeneous structure. In this structure the two sorts of atoms may occupy the available sites at random; they are said to form a *solid solution* (see p. 141). The lattice varies in dimensions as the composition varies, and from results so obtained the composition of any alloy can be determined. This method, combined with the identification method, has been of enormous use in investigating the structure of alloys, and it is true to say that the powder photograph has revolutionized the study of the metallic state.

Thirdly, defects in powder photographs can also be informative. If, for example, the powder is too coarse, it will give 'spotty' lines instead of smooth ones. By counting the number of spots in a line it is possible to estimate the size of the crystals of which the specimen is composed. At the other end of the scale, if the powder is too fine, the lines will be broadened.

Smallness of the crystals is not the only cause of broadening; imperfections will produce the same effect, just as badly ruled gratings will do. Again, the nature and magnitude of the imperfections can be investigated by measuring the broadening of the different orders of diffraction. This method has been of particular use in investigating the way that metals deform when they are bent, drawn or rolled.

Finally, some materials give X-ray photographs intermediate between powder photographs and single-crystal photographs, indicating that the crystals of which it is composed are not distributed entirely at random. This fact may be important in deciding the properties of rolled metal sheets and drawn wires, and the investigations have proved to be of industrial importance. Here the normal powder camera is not of much use, and the longer cylindrical camera employed for single-crystal photographs (fig. 5.4) is of much greater value.

It will thus be seen that the powder method, although much less use for crystal-structure determination, has been far more influential in the world outside academic laboratories. It is for this reason that we have thought it worth while giving so much space to it in this chapter, and to discussing these applications in detail in chapter 12.

CHAPTER 6

how some simple structures were determined

6.1 *Introduction*

In this present chapter we wish to explain how some simple structures were determined from the information given by their diffraction patterns, supplemented by whatever other data were available. Some of these researches seem simple now, but we must remember that, when they were carried out, nothing was known about the sizes of atoms or ways in which they behave in forming crystals; to begin from zero is always difficult.

We shall therefore start with the first crystal structure, which was necessary in order to fix the scale of wavelengths of X-rays (§ 2.8), on which all subsequent work had to be based. We shall then show how structures of increasing complexity could be tackled successfully as the rules of crystal formation unfolded themselves. One could guess how the atoms were likely to dispose themselves, and compare the resultant calculated diffraction patterns with those observed; in this way continuous and rapid progress was made in adding to the store of crystal-structure knowledge.

6.2 *The first crystal structure, NaCl*

The first crystal structure was worked out by W. H. and W. L. Bragg within a year of the discovery of X-ray diffraction. Remember that they had no experience to guide them, that they had to design their own apparatus and superintend its construction, and that the scale of X-ray wavelengths was unknown. The structure that they determined is extremely simple, but this fact must not blind us to the Braggs' remarkable insight. Here we shall try to trace the paths by which they arrived at their result.

They first tried Laue photographs. Apart from verifying that the crystal had cubic symmetry, this method was not particularly productive. The cubic symmetry was evident from the excellent cleavage (§ 3.1) of the crystals, since it is unlikely that a crystal of any other symmetry would cleave so beautifully in three perpendicular directions; Laue photographs taken with the X-rays parallel to the cube edge showed clear four-fold symmetry. Methods of indexing were successful, but they did not provide enough information to solve the problem.

So they then turned to the ionization spectrometer (p. 21) and with this they obtained the results shown in fig. 6.1. This figure contains results for KCl and NaCl, but we shall deal with the latter first, since the former introduces an effect that can be understood only when the structure of NaCl is known.

Let us look at the results for the (100) faces since these show three sets of peaks in pairs which we can recognize as a relatively strong α peak accompanied by a much weaker β peak (p. 22) at a lower angle. The abscissae are values of 2θ, and thus we can say that the values of θ

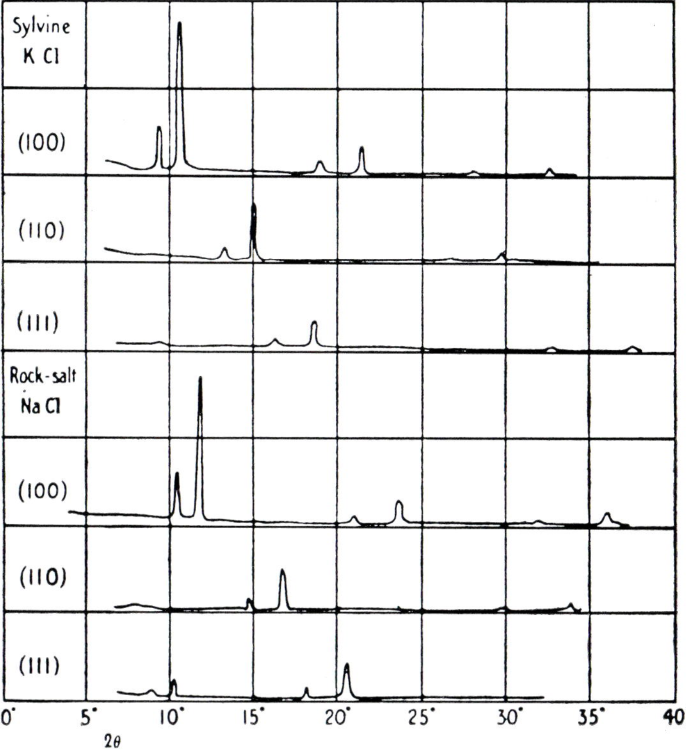

Fig. 6.1. Ionization-spectrometer measurements made by the Braggs for KCl and NaCl.

for the three α peaks shown are 6°, $11\frac{3}{4}$° and $18\frac{1}{2}$°. The sines of these angles are 0·104, 0·203, 0·317. These values are roughly in the ratio 1:2:3, and thus fit in with Bragg's law, giving approximately $\lambda = 2d_{100} \cdot 0{\cdot}104$, from which $d_{100} = 4{\cdot}81\lambda$. The corresponding Bragg angles for (110) are $8\frac{1}{2}$° and 17°, giving $d_{110} = \lambda/0{\cdot}294$, or $d_{110} = 3{\cdot}40\lambda$. Finally the angles for (111) are $5\frac{1}{2}$° and $10\frac{1}{2}$° giving: $d_{111} = \lambda/0{\cdot}186$, or $d_{110} = 5{\cdot}38\lambda$.

What relationships should we expect from these results? As we can see from fig. 6·2, the spacing of the planes (100), (110) and (111) should be in the ratio of $1 : 1/\sqrt{2} : 1/\sqrt{3}$, since they are respectively equal to the

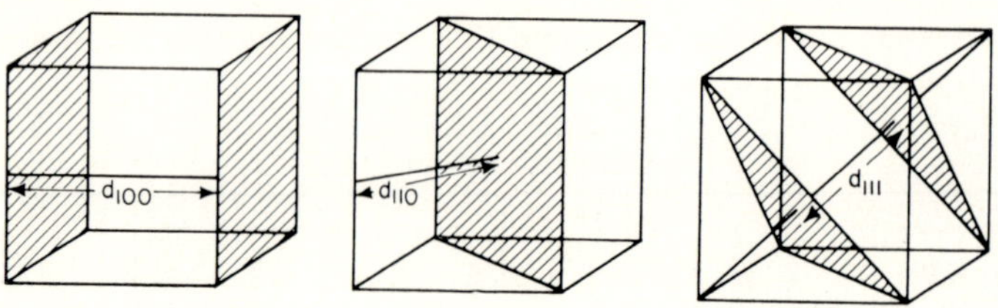

Fig. 6.2. The spacings d_{100}, d_{110} and d_{111} of a simple cubic structure.

cube edge, half the cube-face diagonal, and a third of the cube diagonal from one corner to its opposite. Here was the first difficulty: 3·40/4·81 is equal to 0·71, which is indeed close to $1/\sqrt{2}$, but 5·38/4·81 is equal to 1·12 which is not equal to $1/\sqrt{3}$ but is close to $2/\sqrt{3}$. What is the explanation?

It looks as though something is wrong with the spacings. Are d_{100} and d_{110} too small or is d_{111} too large? We can arrive at the answer by considering the theory of the diffraction grating. The spacing d is the distance between successive lines, which should be all the same. But suppose that, by some mischance, alternate lines are misplaced as shown in fig. 6.3; then the spacing, being the distance between identical lines, is equal to $2d$. But if we explore the diffraction pattern

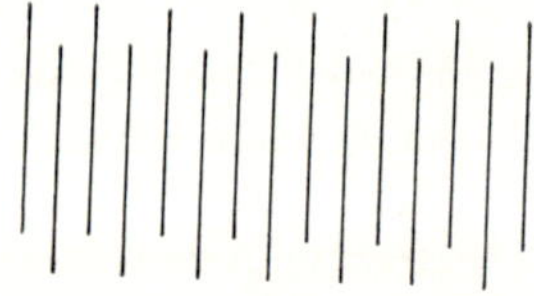

Fig. 6.3. Diffraction grating with alternate lines displaced.

with an ordinary spectrometer, we should not find any evidence of this doubled spacing; we should have to tilt the telescope upwards to find the additional spectra produced. This state of affairs can happen with a crystal. If successive planes contain the same distribution of atoms, the spacing observed will be the perpendicular distance between them; but the spacing may be a mutiple of this if the layers of atoms are displaced with respect to each other.

The problem, then, that faced the Braggs was to find an arrangement of atoms that would give these effects. They were helped by a theoretical paper that had been written in 1896 by Barlow; in this he published his ideas of how atoms might pack together in elements and simple chemical compounds. It will be noted that this paper was published long before there was any hope of verifying his ideas experimentally, and it is likely that nowadays such a paper would not be accepted; the editors of scientific journals prefer papers with practical support for any ideas that are contained in them.

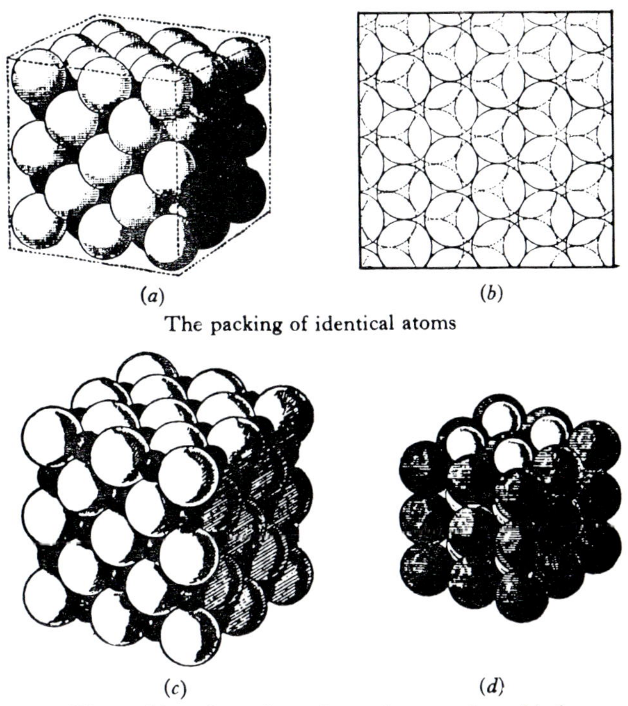

Fig. 6.4. Some possible atomic arrangements suggested by Barlow. (*a*) and (*b*) packing of identical atoms; (*c*) and (*d*) packing of two sorts of atoms.

Four diagrams from Barlow's paper are shown in fig. 6.4. The third one (*c*) satisfies the conditions deduced from the spacings of NaCl. Figure 6.5 shows that successive (100) and (110) planes contain identical arrangements of Na and Cl atoms, but the (111) planes contain sodium atoms and chlorine atoms alternately. Thus the observed (111) spacing is the true one—the distance between planes of similar atoms; but the (100) and (110) spacings are only half those calculated from the dimensions of the unit cell.

There is another piece of evidence that should support these deductions. If the atoms in successive (111) planes scattered X-rays equally, the spacing again would appear to be halved, and the first-order spectrum would not appear. If the atoms are different, the structure factor should be small and this is what we observe in fig. 6.1; the first-order (111) reflection from NaCl is weaker than the second order. The third order is too weak to detect. In KCl, the atoms contain nearly equal numbers of electrons (19 and 17) and so the first- and third-order reflections from the (111) planes cannot be detected. This is the anomaly that was mentioned earlier in this section.

So the Braggs derived the structure of NaCl and KCl, and the start of a great subject had been made. But it should be noted that they had derived only the *arrangements* of atoms; the scale was expressed in terms of the wavelength of the K radiation of palladium and this was unknown. However, it could now be found, for the number of atoms in the unit cell of NaCl was the last item of information needed to find the size of this unit cell.

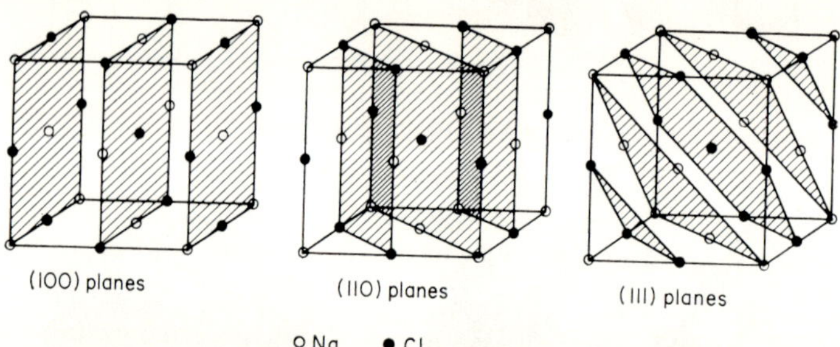

Fig. 6.5. Unit cell of NaCl structure, showing that (100) and (110) planes contain mixtures of Na and Cl atoms, but (111) planes consist of alternating layers of Na and Cl.

6.3 *Determination of X-ray wavelengths*

The unit cell of NaCl contains four Na and four Cl atoms. (Do not be misled into thinking that a drawing such as that in fig. 6.4 indicates more; remember that an atom at a corner is shared by eight unit cells, an atom on an edge is shared by four, and an atom on a face is shared by two.) The mass of the unit cell is therefore obtained by dividing the sum of the atomic weights of the atoms by Avogadro's number— $6{\cdot}025 \times 10^{26}$ atoms per kg-atom. The mass, however, is also the volume multiplied by the density, and since the cell dimension is the only unknown, this can be derived. Let us put this idea into figures.

The mass of the atoms is $4(23+35{\cdot}5)/6{\cdot}025 \times 10^{26}$ kg.

The density of rock salt is $2{\cdot}16 \times 10^{3}$ kg m^{-3}. Thus

$$a^3 = \frac{4 \times 58{\cdot}5}{2{\cdot}16 \times 10^3 \times 6{\cdot}025 \times 10^{26}}, \text{ whence } a = 5{\cdot}64 \times 10^{-10} \text{ m.}$$

This was the first atomic dimension derived. (The Braggs obtained $5{\cdot}628 \times 10^{-10}$ m.) The unit 10^{-10} m is an extremely useful one for expressing interatomic distances and is given the name Ångstrom unit (Å) after a famous Swedish spectroscopist who used it for expressing wavelengths of spectral lines. It is equal to 0·1 nm.

From this value of a, we can derive the wavelength of Pd Kα radiation; it turns out to be $(5{\cdot}64/2 \times 4{\cdot}81)$ Å from the result on p. 79). This is 0·59 Å. (The accepted value now is 0·587 Å.) This value laid the foundations of a new subject—X-ray spectroscopy—in which considerable accuracy, of the order of 1 part in 100 000, has now been achieved. It was soon found that rock salt was not a good standard since perfect crystals were not easily obtained. But the story of the adoption of new standards, and the final production of X-ray wavelengths by diffraction from ruled gratings is too long and detailed to be included in this book.

So the stage is now set. We have X-rays of known wavelength and we have established that the diffraction patterns of crystals can be explained in terms of arrangements of discrete particles. We must not forget that this result was not inevitable; it could have turned out that atoms had some static infra-structure that would have led to much more complicated results. But simplicity prevailed, and W. L. Bragg and others went on to determine more structures of increasing complexity.

6.4 *Diamond and iron pyrites*

We have chosen two more cubic crystals to illustrate the progress of the subject, but we do not propose to describe the steps in these investigations in as much detail as we used in the last section; the determination of the structure of NaCl was a unique step which deserved a full section to itself.

Diamond was one of the next crystals studied by W. L. Bragg. It was an obvious choice, being the most precious crystal of all, and, since it belongs to the cubic system, it was likely to yield to the same methods that were successful for NaCl. Preliminary work showed that the unit cell, of edge 3.6 Å, contained eight carbon atoms.

Now this is the same number of atoms as for NaCl. Can we therefore use the same arrangement, but replacing all the Na and Cl atoms by C? Clearly we cannot, for fig. 6.5 shows that if all the atoms were the same, the unit cell edge would be halved and the atoms would be arranged on a simple cubic lattice. This result is obviously wrong.

If we try the same approach as for NaCl, we find the following results:

$$d_{100} = \frac{1}{4}a_0; \quad d_{110} = \frac{1}{2\sqrt{2}}a_0; \quad d_{111} = \frac{1}{\sqrt{3}}a_0.$$

The first result means that the atoms must be arranged in planes dividing the cell edges into four parts, but the arrangements must differ in some way for each of the four planes. We may start by using the Na arrangement in NaCl for half of the C atoms, and then interpolating atoms between them arranged as shown in fig. 6.6. We can see that this pattern of atoms gives the correct relative spacings for d_{100} and d_{110}; a three-dimensional model is needed to show that d_{110} is also correct.

This is the right answer. It is sometimes called the diamond lattice, but this is incorrect. The atoms are not arranged on a lattice because, as we explained on p. 32, lattice points must have exactly the same environment *in the same orientation*; in diamond all the atoms have a

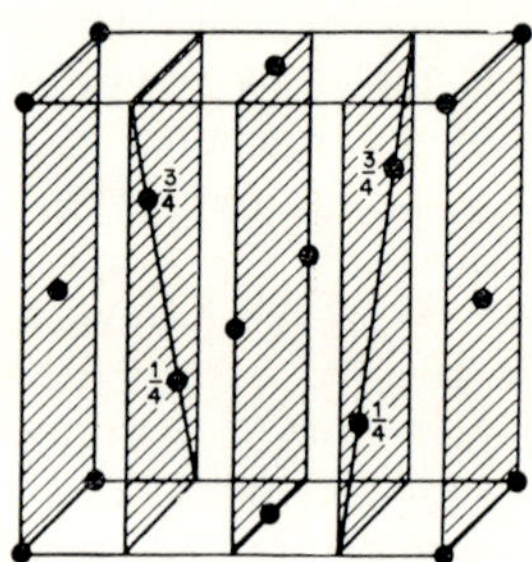

Fig. 6.6. Unit cell of diamond structure showing distributions of atoms in successive layers parallel to (100).

tetrahedral arrangement of atoms around them, but for some—such as those at height $\frac{1}{4}$ and $\frac{3}{4}$ in fig. 6.6—the tetrahedra are inverted with respect to those around atoms at height 0 and $\frac{1}{2}$.

Iron pyrites, FeS_2, was another crystal that soon yielded to an assault by W. L. Bragg. There is an important reason for including it here; it was the first structure in which the atoms were not fixed by symmetry; the S atoms lay in positions that are defined by a variable parameter, and the value of this parameter had to be determined in order to find precisely the positions of the atoms in the unit cell. Crystals with large numbers of variable parameters are dealt with successfully now but it is important to recall the times when the existence of even one such parameter was regarded as presenting a formidable problem.

The unit cell, of side 5·4 Å, is cubic and contains 4{FeS_2}. We can arrange the four Fe atoms in the Na positions of NaCl, but clearly the eight S atoms cannot be located by analogy with either NaCl or diamond. Bragg decided that the S atoms lay in special positions on the three-fold axes (see § 6.6), and thus some way must be found of determining how far along these axes they were placed. Two atoms must be associated with each of the four Fe atoms, and since these are in known positions one parameter will place all the eight S atoms in the unit cell.

Now the three-fold axes of a cube lie along the diagonals—that is, from one corner to the opposite corner—but Bragg found it impossible to find positions for the sulphur atoms that would give the observed intensities of the orders of diffraction. There was something wrong. This sort of problem often occurs in science; one has tried to be quite logical in one's assumptions, but the results are not self-consistent. It is quite difficult to find out which step is wrong. For FeS_2, the wrong step was to assume that the three-fold axes were *along* the cube diagonals; they need only be *parallel* to them, and can form a non-intersecting set as shown in fig. 6.7. As soon as Bragg realized this, the correct structure became apparent.

Nowadays the difficulty would not occur. The theory of space groups (p. 38) would have indicated immediately what arrangements of three-fold axes were possible. But in 1913, this theory was not yet

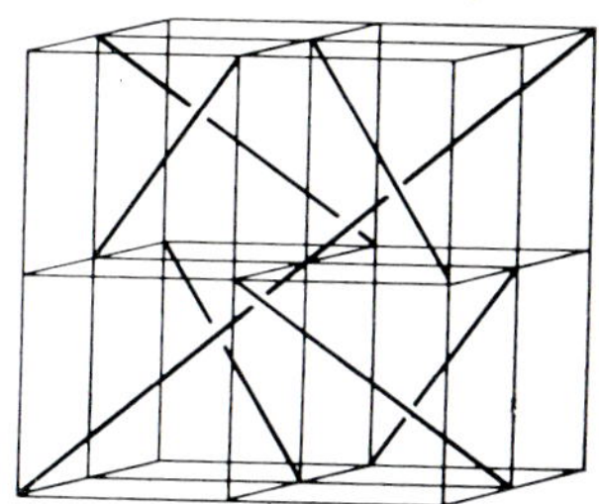

Fig. 6.7. Cube, showing non-intersecting three-fold axes. Each of the lines lies along a diagonal of one of the eight small cubes into which the unit cell is divided.

fully appreciated, and Bragg's deduction had to be made from first principles. This fact adds still more distinction to this pioneering piece of research.

The fact that a variable parameter is involved is shown by the irregular variation of the intensities of the successive orders of diffraction; there are no simple sequences as there are for NaCl and diamond. Thus for the (100) planes, the 200 reflection is strong, 400 and 600 are weak and 800 is stronger again. For the (111) planes, 555 is outstandingly strong. The 800 reflection suggests that the parameter u is near to $\frac{1}{8}$, $\frac{3}{8}$, $\frac{5}{8}$ or $\frac{7}{8}$; then the S atoms will scatter in phase with the Fe atoms for this reflection. The value cannot be near to $\frac{1}{4}$ or $\frac{3}{4}$; although these values would also make 500 strong, 400 would be strong also.

The problem is to select the right value from the four possibilities. We can reduce them to two by noting that $\frac{7}{8}$ can be expressed as $-\frac{1}{8}$, since we can take any corner of the unit cell as origin; because the atoms are arranged in pairs equidistant from the origin, we need not consider both $\frac{1}{8}$ and $\frac{7}{8}$, nor both $\frac{3}{8}$ and $\frac{5}{8}$. We must decide only between $\frac{1}{8}$ and $\frac{3}{8}$.

The decision can be made by referring to the reflection 555. Since this is strong, we must choose a value of u that makes $5u$ near to a whole number, so that the atoms scatter in phase; if it is near to a whole number plus a half, the S atoms will scatter out of phase with the Fe atoms and the reflection will be weak. Now $5 \times \frac{1}{8}$ is near to $\frac{1}{2}$, but $5 \times \frac{3}{8}$

is near to 2. Thus $\frac{3}{8}$ is the right answer. This is not a fixed fraction however. More careful consideration gave the value of u as 0·386. (Bragg's original paper gave 0·114, but he chose a different origin from that used here.)

It must be pointed out that this is an over-simplified description of the structure determination; the conditions for reinforcement of the scat-

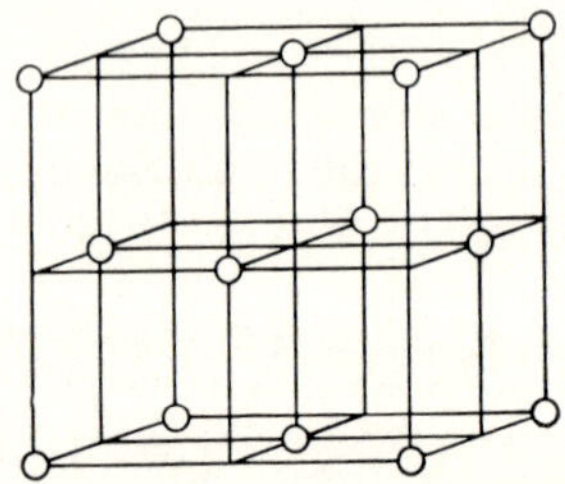

Fig. 6.8. The face-centred cubic structure.

tered waves for the 555 reflection are really more complicated than we have indicated; but the basic conclusion is sound.

6.5 *Results from powder photographs*

Not all the early results were obtained by single-crystal methods; the powder method (§ 5.9) also allowed a large number of polycrystalline materials to be studied. Metals and alloys, for example, were found to be crystalline and some of the simple structures could be determined from the powder patterns alone. The crystal structures of many of the elements were found in this way, and added greatly to our store of knowledge of the solid state.

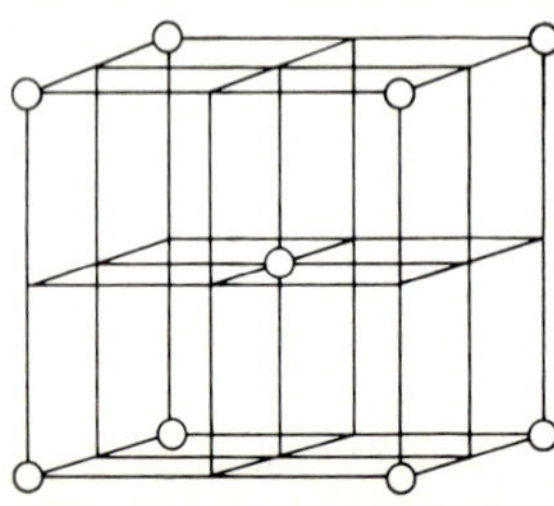

Fig. 6.9. The body-centred cubic structure.

The reason why information could be obtained in this way is that many of the structures are very simple and therefore their patterns were easily recognizable (fig. 5.13). Two cubic structures turned out to be particularly common; they are called the face-centered cubic (fig. 6.8) and body-centred cubic (fig. 6.9) because the atoms lie on the points of these two lattices. Again it should be noted that this usage is not strictly correct; a structure can have a face-centred or body-centred cubic lattice and have a large number of atoms in the unit cell. A better name for the face-centred structure is cubic close-packed; this name indicates that the arrangement of atoms is formed by packing spherical

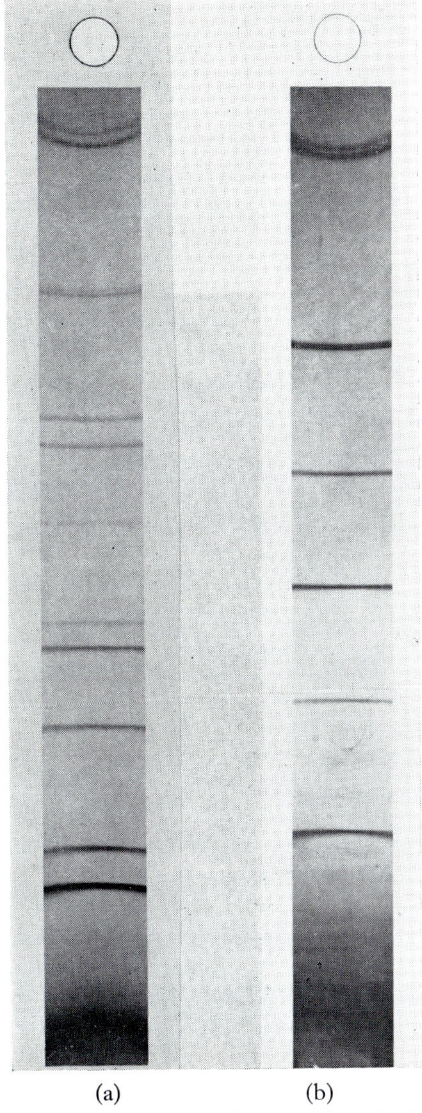

(a) (b)

Fig. 6.10. Powder photographs of the structures shown in figs. 6.8 and 6.9. (*a*) face-centred cubic, (*b*) body-centred cubic.

atoms as closely as possible together in such a way that the unit cell is cubic. It then transpires that the atoms lie on the points of the face-centred cubic lattice (fig. 6.8). The so-called body-centred cubic structure has two atoms in the unit cell—one at the corner and the other in the centre. This is not quite close-packed but is a fairly good approximation.

The two sorts of structure give powder patterns that are readily recognizable (fig. 6.10). Different elements have unit cells of different sizes, and so the scales of the patterns will be different, but the *sequences* of the lines are quite characteristic. It will be noted, however, that the intensities of the lines vary in a rather odd way, and to account for this variation we have to introduce a factor that we did not mention in the section on powder photographs in the last chapter. It is called the *multiplicity factor*, and is simply a measure of the number of sets of planes that can contribute to a particular powder line.

For example, six sets of planes contribute to the 100 line, corresponding to the six faces of a cube; the corresponding reflections have the indices 100, 010, 001, $\bar{1}00$, $0\bar{1}0$, $00\bar{1}$. To find the multiplicity factor for a cubic crystal we transpose the indices in all possible sequences, and also include both positive and negative values. Thus the line 100 has only a small multiplicity factor, but for a more general line, such as 321, the number of possible combinations is much larger—48. (The reader may try to deduce this value for himself; if the exercise does nothing else it will teach him to be systematic!) Thus lines with general indices appear much stronger than those with special indices such as $h00$, $hh0$ or hhh.

There is a third important type of powder photograph which corresponds to another form of close-packing; this is called the hexagonal close-packed structure. Its relationship to the cubic close-packed structure can be explained by considering the way in which we should try to build a close-packed arrangement of spheres. We should first build a close-packed plane (fig. 6.11) and then place another close-packed plane upon it. Another plane of spheres would then be placed upon the second. But for this plane we face a dilemma; there are two essentially different places into which it can be put so that each sphere rests in a depression in the plane below. In one, each atom is directly above an atom in the first plane; this is hexagonal close-packing. In the other, the atoms occupy positions that are not directly above those in the first plane; these are alternative sites that they can occupy (fig. 6.11). The next layer then occupies positions directly above the first. This arrangement is cubic close-packing.

Of course, any sequence of planes that fits closely together is close-packed, but it is remarkable that nature chooses one or the other, and only in one element—cobalt—does it show any doubt. It is also odd that elements that choose the close-packed hexagonal structure behave as if their atoms are slightly flat, except for two elements—zinc and cadmium—for which they behave as though they were very elongated. The causes of these discrepancies are not fully understood.

Apart from this application to these simple structures, powder photographs have not been much used for structure determination. It is rather surprising therefore that one very important structure was determined in this way as early as 1926; this was an alloy called γ-brass, Cu_5Zn_8. It contains more zinc than ordinary brass, and is extremely

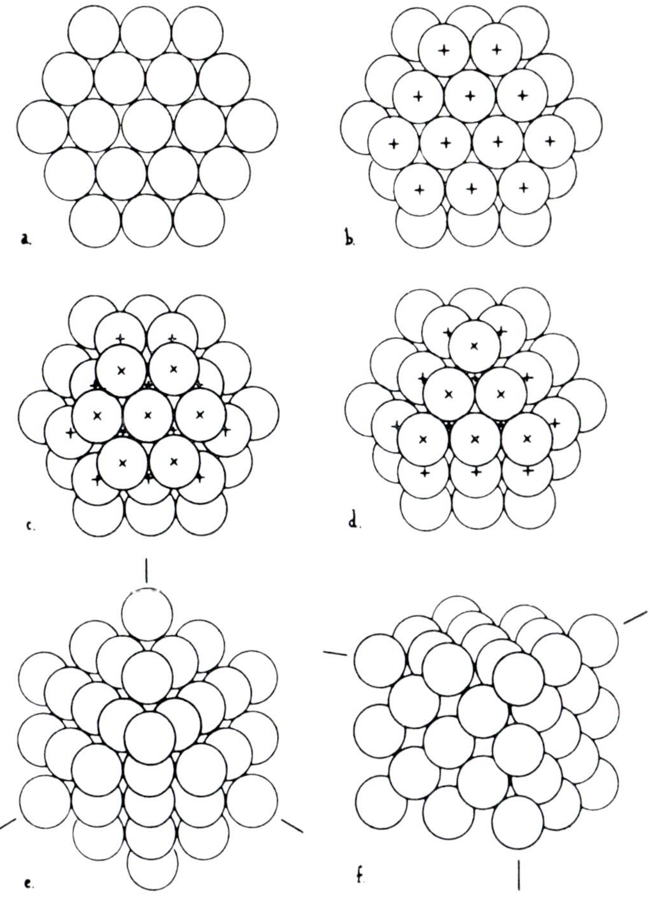

Fig. 6.11. (*a*) Close-packing of one layer of atoms; (*b*) a second layer; (*c*) and (*d*) alternative positions for the third layer; (*e*) extension of (*d*); (*f*) another view of (*e*) to show that the structure is cubic.

brittle, so that it has no practical importance. But its structure played a large part in developing the theory of the metallic state, as we shall show in Chapter 10. The story of the investigation is remarkable for its simplicity and directness.

Bradley, working under W. L. Bragg in Manchester, had developed a peculiar understanding of powder photographs and, with Thewlis, decided to attempt the structure, which seemed too complicated for its time. It was body-centred cubic and its unit cell of side 8·85 Å contained 52 atoms—20 Cu and 32 Zn. Success was made possible by the fact that the powder photograph showed a resemblance to that of an ordinary body-centred cubic structure with $a = 2{\cdot}95$ Å; the strong lines of the pattern lay in the places where the lines from this structure would occur.

Bradley and Thewlis therefore suggested that, to a first approximation, the unit cell of the γ-structure could be formed by stacking together 27 small cubes (fig. 6.12). These would contain 54 atoms, whereas we noted that the unit cell contained only 52; which two should we remove? We also noted that structure was body-centred. The 27 small cubes do form a body-centred unit cell, and will continue to do so if we remove the atoms at the corner and at the centre of the centre

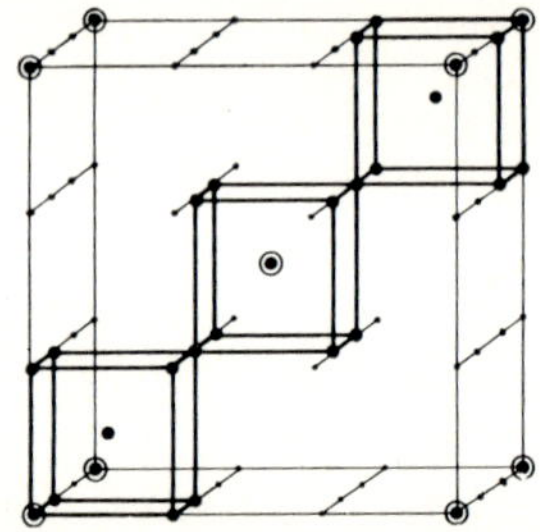

Fig. 6.12. The derivation of the structure of γ-brass from the body-centred cubic structure. The large cube is composed of 27 small cubes, three of which are shown, heavily outlined. The small spots indicate the corners of the small unit cells and the large spots indicate atomic positions in them. The atoms that are ringed are those that are removed.

cube (fig. 6.12). We now have a structure with the right unit cell and it is also body-centred.

But it has two large spaces in it where the two atoms have been removed. Bradley and Thewlis therefore shifted the remaining atoms about so that they tended to fill in these spaces. After a certain amount of trial, they produced a set of atomic positions, consistent with the cubic symmetry, that gave excellent agreement between calculated and observed intensities. The complete structure was based upon five variable parameters. To have solved such a structure, from powder photographs alone, in 1926, was a remarkable achievement indeed.

In fact, no one of Bradley's abilities has emerged since he retired from his work. He was responsible for determining a number of alloy structures that could be regarded as slight departures from the sample structures described earlier. He and Thewlis also worked out the structure of one form of manganese—a complicated atomic arrangement for an element to have—and this has given rise to some ideas of atomic structure that will be discussed in Chapter 10. It is also remarkable that in quite recent attempts to improve the accuracy of Bradley's work, no significant inaccuracies have been found in any of it.

It will be seen, therefore, that although powder methods made considerable contributions in the early days of the subject, they cannot be regarded as being in the main line of development. They were an interesting side-line and we must now revert to a further discussion of single-crystal methods.

6.6 *Influence of symmetry theory*

We have seen how the first crystal structures were derived without the help of symmetry theory; so long as a structure gave the right diffraction pattern it was regarded as correct. As the importance of symmetry began to be realized—mainly under the influence of Wyckoff, who produced a most valuable set of tables of symmetry-related points (p. 38)—it became usual to check that a structure did obey the space-group rules; it then became clear that some proposed structures were wrong. In present-day work, every structure investigation starts with a space-group determination, which may lead to the true space group uniquely, or to a small number of possibilities.

As an example, let us consider a structure that has fascinated crystallographers over the centuries—the alums. These have considerable commercial importance, they form beautiful crystals, and there is an immense variety of chemical composition. The basic formula is $KAl(SO_4)_2 . 12H_2O$; but K can be replaced by other monovalent elements and by NH_4, the Al can be replaced by the trivalent elements Fe and Cr, and the S can be replaced by Se. The crystals are cubic and the unit cell was soon found to have the relatively large value of 12 Å. Several early workers tried to find the structure and put forward incorrect results based on inadequate evidence. Let us assess the complexity of the problem.

The unit cell of potassium aluminium alum contains $4\{KAl(SO_4)_2 . 12H_2O\}$. Symmetry considerations place the K and Al in an NaCl arrangement (p. 82) and so these atoms are fixed with no parameters. The S atoms have the same symmetrical arrangement as in FeS_2, so that one parameter will fix them all. Of the oxygen atoms one from each sulphate group must also lie on the three-fold axis and so another parameter is required; the other oxygen atoms lie in general positions (p. 57) and thus three parameters are needed to define them. The water molecules—each of which can be regarded as a single scatterer since the scattering by hydrogen is negligible—must be divided into two sets of 24, and are thus fixed by six more parameters. The total number of variable parameters to define the complete structure is therefore 11.

This is a very small number by present-day standards, but in the

1920's it was too large to handle. It has been said that the dividing line between arts and science is the number six: with less than six parameters to be handled at once, systematic methods can be used; with more than six, guesswork and intuition—which are usually associated with the so-called humanities—have to be brought into play. However true this generalization might be, the alums certainly did not succumb to the traditional methods, and had to await some of the further development that will be described in Chapter 9.

Space-group symmetry can give other types of information as well. For example, if there is a two-fold axis in a structure, there must be a region around it that cannot contain any atoms, except those that lie precisely on it (p. 39). No atom can lie within a distance of its atomic radius from the axis, since it must not overlap with its symmetry-related neighbour. The same rule applies to planes of symmetry, there must be layers on either side of these planes on which atoms cannot lie. These and other rules were summarized by Bragg and West in 1928, in a paper entitled 'A technique for the X-ray examination of crystal structures with many parameters'. The word many—meaning 20 or 30—reads rather amusingly now, when structures with thousands of parameters have been successfully solved (see § 8.10).

An excellent example of the methods described by Bragg and West is given by the crystal beryl, a silicate of beryllium and aluminium. It is a precious stone used in jewellery, the most well-known form being emerald; the green colour is caused by slight traces of other elements that can replace the metal atoms. Bragg and West found that the unit cell, which was hexagonal, contained $2\{Be_3Al_2Si_6O_{18}\}$. This looks a complicated problem, yet, in fact, it proved to be absurdly easy and was solved in a single afternoon!

The reason for the simplicity was that the space group that was found contained a large number of rotation axes—6-fold, 3-fold and 2-fold—and reflection planes. As we have explained, each of these symmetry elements gives spaces in which atoms of a given size cannot lie. It turned out that there were two-fold axes at distances of 1·29 Å from the mirror planes; since oxygen was known to have a diameter of 2·7 Å its possible positions were very severely limited. Of the 36 oxygen atoms in the unit cell, 12 had to lie in special positions, and it was soon found that they could not lie on the two-fold axes because then they would so greatly limit the positions of the other atoms that none was possible at all; they therefore had to lie on reflection planes.

This simple fact proved to be sufficient to define the positions of all the oxygen atoms; there was only one way in which they could pack into the unit cell consistent with the known symmetry. Since there were only four aluminium atoms in the unit cell, these must lie in very symmetrical positions; since it was known that they tend to lie in places surrounded by six oxygen atoms—that is, they have six-fold coordination—their positions were soon found. The same is true of the six beryllium atoms;

these have four-fold coordination; this fact is the basis of all silicate structures. Thus the position could easily be found. The result was checked against the observed X-ray intensities, and was found to be correct, requiring only slight modification. There are not many structures that can be worked out so delightfully.

A quite different example is given by another silicate, cyanite. This is triclinic (p. 35) and the unit cell has dimensions $a = 7{\cdot}1$ Å, $n = 7{\cdot}7$ Å, $c = 5{\cdot}6$ Å, $\alpha = 90°$, $\beta = 101°$, $\gamma = 106°$, it contains $4(Al_2SiO_5)$. The crystal is centrosymmetrical (p. 35) and therefore 48 parameters are needed to define the positions of the atoms in the unit cell. It must have taken considerable courage to tackle a problem of this sort in 1928, but it was successfully solved by Taylor and Jackson in Manchester.

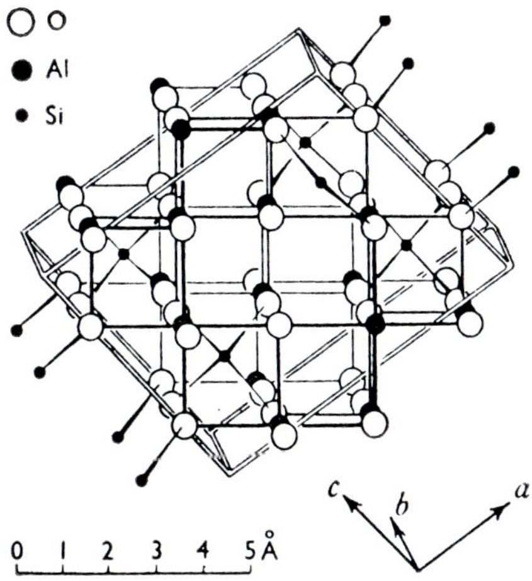

Fig. 6.13. How the triclinic unit cell of cyanite is derived from the cubic close-packed lattice.

The first evidence was that the crystal had a very high refractive index, 1·72. It was known that this property was associated with close-packing of atoms, and it was therefore probable that the oxygen atoms were packed tightly together in the unit cell. This conclusion was supported by working out the volume associated with each oxygen atom; it came to 13·6 $Å^3$, almost identical with that produced by cubic and hexagonal close-packing. The aluminium and silicon atoms are so small that they will fit in the spaces between the oxygen atoms.

But how can we explain the odd shape of the unit cell? Here three-dimensional geometry comes into play. If we take the cubic close packed structure (p. 89), it is possible to define a unit cell by taking any one atom and joining it to any three noncoplanar atoms. Can we find a triclinic cell of the known dimensions of cyanite?

The oxygen atom has a radius r of 1·35 Å; this was known from simpler structures. A cubic close-packed arrangement of atoms of this size would have a cell dimension of $2\sqrt{2}r$ (fig. 6.4 *a*), which is equal to 3·82 Å. Let us call this d. Now Taylor and Jackson found that the cell edges of cyanite could be expressed pretty accurately as $\sqrt{(7/2)}d$, $2d$ and $\sqrt{2}d$; these correspond to the distance between atoms shown in fig. 6.13. The axes b and c are clearly perpendicular, since c is in the plane of the diagram and b runs directly away; this accounts for the value $\alpha = 90°$. The axis can be chosen to give close approximation to β and γ. This suggestion proved to be correct. Positions for the aluminium and silicon atoms could be found with the right coordination, and so this difficult problem was solved without recourse to more complicated methods.

6.7 *Conclusions*

The examples in this chapter show some of the ways in which crystal-structure problems were solved up to about 1930. There were no general methods; each crystal had to be considered as a problem on its own and ways had to be found of circumventing the difficulties it presented. The most successful workers were those with the most adaptable minds, who could see the possibilities of an unusual approach. As more and more structures were solved, so more and more information became available to help with fresh problems.

Obviously the subject could not continue in this way; the existence of new ways of approaching problems could not be guaranteed. A structure like cyanite, with 48 parameters, might be possible because of some unusual property, but in general the handling of more than 20 or 30 parameters was too big a task. The paper by Bragg and West, which was supposed to be a guide to other workers, can now be seen to be of historical interest only, because at the same time a new and powerful approach was developing. This was the Fourier method, based on an idea of W. H. Bragg, and developed by W. L. Bragg in a famous paper published in 1929. The next three chapters will be devoted to explaining the method, to indicating its physical basis, and to describing some of the results obtained with it.

CHAPTER 7
Fourier methods

7.1 *Explanation of Fourier series*

FOURIER is one of the great names in mathematical physics. J. B. J. Fourier was one of Napoleon's great scientific advisers, and he was very much concerned with the application of mathematical methods to the solution of practical problems. His particular interest lay in heat transmission and he wanted to find out, for example, how the periodic application of a temperature change to the end of a conducting bar would be transmitted along the bar. He could solve the problem if the temperature change were sinusoidal, but not if it were discontinuous.

He therefore conceived the idea of trying to express the discontinuous change as the sum of a number of sinusoidal changes: each one could be considered separately; the heat waves would presumably travel with different speeds and the results could be added together at any point in the bar. Thus a general solution should be possible.

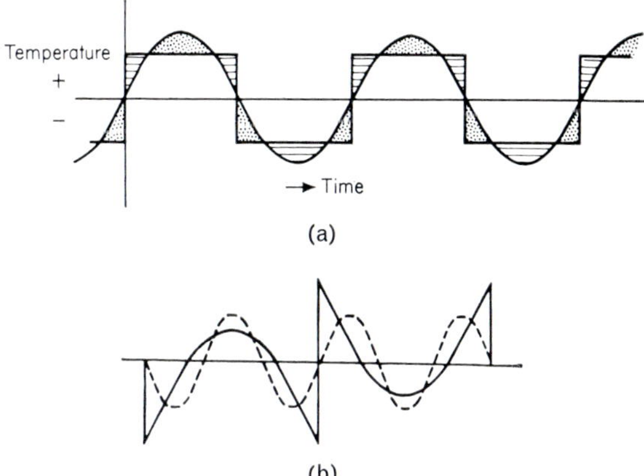

Fig. 7.1 (*a*) Square wave of temperature against time, showing approximate representation by sine curve; (*b*) curve formed by shaded areas of (*a*), showing that it can be approximately represented by a sine curve of one-third wavelength (not to same scale as (*a*)).

It is easy to see how the method works. Suppose that the temperature of the end of the bar is suddenly increased and then suddenly decreased at equal intervals, to give what is known as a square wave (fig. 7.1 *a*). To a first approximation, this can be expressed as a sinusoidal wave, whose amplitude we can adjust so that the difference between the two curves is equally positive and negative; the total dotted and hatched areas are equal for each half of the sine wave. If we plot these differences, we obtain a more complicated curve (fig. 7.1 *b*), which is on the whole nearer to the mean line. This can be expressed roughly as a sine curve of one-third of the wavelength, shown by the broken line in fig 7.1 *b*. By continuing this process, we can reduce the discrepancy

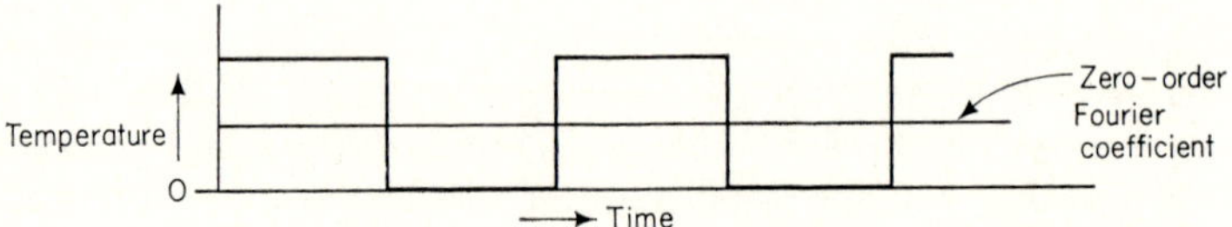

Fig. 7.2. Square wave of fig. 7.1 with different origin, showing derivation of zero-order Fourier coefficient.

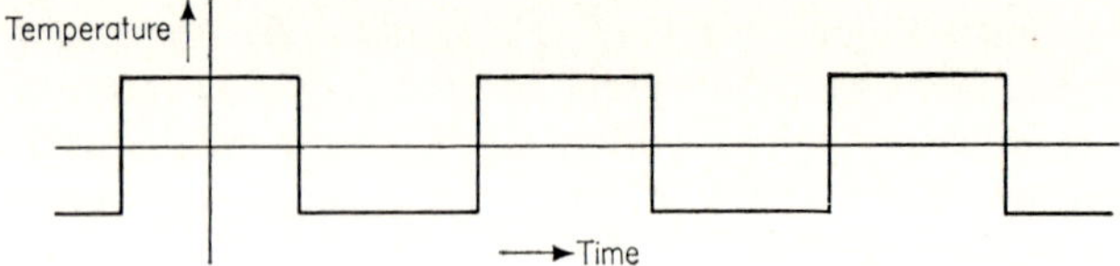

Fig. 7.3. Square wave of fig. 7.1 expressed as an even function.

step by step, and can obtain as close an approximation as we like to the original curve. But we can never reproduce it precisely.

The amplitudes of the various sinusoidal waves are known as the *Fourier coefficients*, each being specified by a single integer. In the example that we have chosen, the first Fourier coefficient is large, the second one is zero (because there is no need to introduce a wave with half the wavelength of the original), and the third one is small. If we

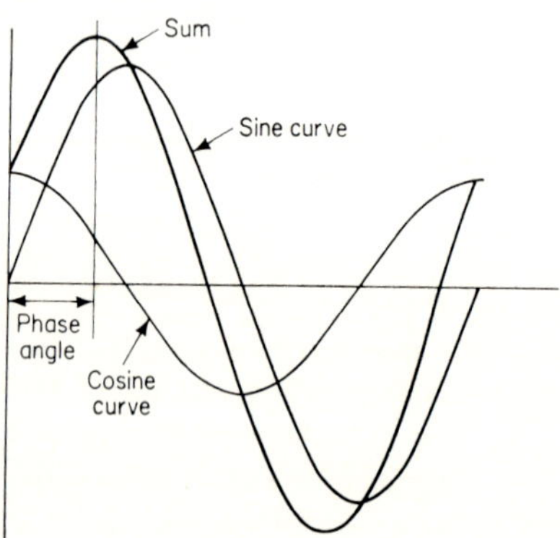

Fig. 7.4. Addition of sine curve and cosine curve showing that the sum is also sinusoidal of same wavelength. The phase angle of the sum curve is indicated.

were to continue the process we should find that all the even coefficients were zero, and the odd ones gradually decreased as the wavelength decreased. This is the particular property of the example chosen; it is not, of course, generally true.

There is also a zero coefficient, corresponding to a sinusoidal curve with infinite wavelength—that is, a constant. In our example, this also has zero magnitude, but if we had taken the zero of our curve at the lowest level (fig. 7.2) it would have been finite. The zero coefficient is a measure of the amount by which the curve exceeds zero on the average: it is proportional to the integral of the function over one cycle. We shall not discuss here the method of finding Fourier coefficients mathematically; not all functions can be dealt with by such methods, but numerical methods or digital computers can be used for any reasonable function, however arbitrary its shape.

The process of finding the Fourier coefficients for a given function is called *Fourier analysis*. Fourier analysis is not unique; that is, we can obtain somewhat different answers for the same curve if we change the way that we describe it. As we have seen, altering the level of the function alters the zero coefficient, but it does not alter any of the others. Altering the origin however has a more significant effect. We have seen that, by taking the origin half-way up one of the vertical lines, all the Fourier components were sine curves. If, however, we had taken the origin a quarter of a wavelength along (fig. 7.3) all the components would have been cosine waves, but with the same amplitudes as the sine waves. If we had taken the origin at some intermediate point, each component would have been partly a sine wave and partly a cosine.

The description of the curve that makes all the components sine curves produces what is called an *odd function*; as we pass through the origin, the ordinates change sign. The second description makes it an *even function*; as we pass through the origin, the ordinates do not change sign. If the function is neither even nor odd, the Fourier components have both sine and cosine parts. But if we add a sine and cosine curve of the same wavelength, we obtain a sinusoidal curve also of the same wavelength, but with a different origin (fig. 7.4). The amount by which this is displaced from the origin for the cosine curve is called the *phase* of the coefficient; it is usually measured as an angle, either in degrees with the complete wavelength as 360° or in radians with the complete wavelength as 2π. The Fourier analysis of a function is specified, with respect to the origin chosen, by the Fourier coefficients each given as an amplitude and a phase or as a combination of cosine and sine parts. (There is an alternative expression in terms of complex exponentials, but we shall not deal with it here although for theoretical work it is the more useful.)

If a function has been analysed into its Fourier components, one can test whether the right answer has been obtained by adding the curves graphically and seeing if the total is in reasonable agreement with the original curve. This process is called *Fourier synthesis*. If the answer is right, the Fourier synthesis should oscillate about the original curve, intersecting it in as many points in each wavelength as the number of Fourier components used (fig. 7.5).

7.2 *Application to X-ray crystallography*

What has all this to do with diffraction of X-rays by crystals? The connection was first seen by W. H. Bragg in 1915, only three years after the discovery of X-ray diffraction. As so often happens in science, the idea was well ahead of its time, and was not developed, as we shall show, for over twenty years.

Bragg was invited to give one of the important lectures to the Royal Society—the Bakerian Lecture—and used the occasion to underline the important work that he and his son had started and also to put forward some more fanciful ideas. This is one of the purposes of such lectures; if one publishes a scientific paper, it is usually concerned with a specific

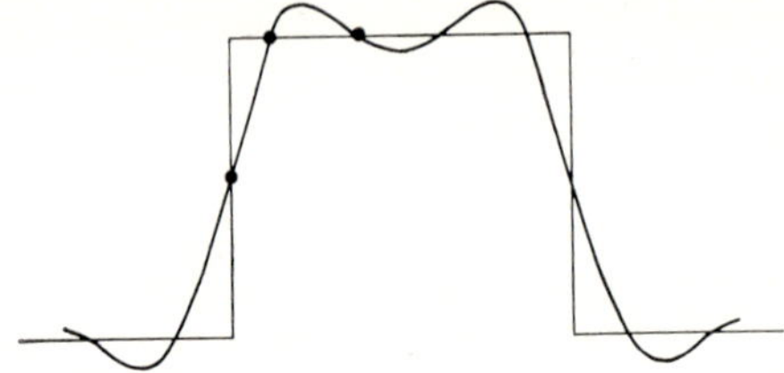

Fig. 7.5. Representation of one period of a square wave by the first three terms of its Fourier series. The three intersections with the square wave are shown by dots.

subject, but if one is invited to give a lecture one can include ideas that are only speculative. (Maxwell's theory of electromagnetic waves was based upon an idea put forward by Faraday when he had to deputize for an absent speaker at the Royal Institution; Faraday was an experimentalist, and would not normally have thought of publishing theoretical ideas.)

The suggestion that Bragg made was that, since a crystal has a periodic structure, it should be representable by a Fourier series. Since the crystal is three-dimensional the Fourier series would also have to be three-dimensional; that is, each coefficient would have to be specified by *three* integers, not just one (cf. p. 50). Now, as we have seen, an X-ray reflection is also specified by three integers. Was there some connection between the two sets? Bragg thought that there was. The process of diffraction, he suggested, was really a process of Fourier analysis: the different orders of diffraction provide the Fourier coefficients of the electron distribution in the crystal; the difficult problem is to find how to derive the Fourier coefficients from the observed intensities.

It was quite impossible, at the time (1915), to develop the idea, and when research was resumed after the First World War enough straightforward problems were available to keep people busy. But Bragg's idea was developed theoretically. It was realized that the Fourier coefficients were the structure factors (p. 58) of the X-ray reflections, and this raised an immediate difficulty.

A structure factor, as we have seen on p. 58, is specified by an amplitude and a phase angle. The amplitude can be derived from the intensity of the reflection, but we have no knowledge of the phase. The relative phases of the waves that reach the X-ray film or the diffractometer counter are lost in the process of recording. Thus there is no way of finding the Fourier coefficients experimentally; we cannot sum the Fourier series and so find the electron distribution in a crystal directly from its X-ray diffraction pattern. This quandary is called the *phase problem.*

Of course, if we know the arrangement of atoms in the unit cell we can calculate the phases of the structure factors, but there does not seem much point in summing a Fourier series merely to tell us what we know already. In fact, however, the Fourier series *does* give us extra information; in addition to telling us where the atoms are it tells us also how the electrons are distributed in them. Some American workers made accurate measurements of the diffraction pattern of rock salt in order to find the electron distribution in the sodium and chlorine atoms.

Apart from the theoretical problem, there is also a practical one—how to sum a three-dimensional series. It can readily be appreciated that summing a one-dimensional series is a lot of work; to carry out the summation in three dimensions would seem to be prohibitively long. Rock salt is a special case; because its unit cell is small it gives very few reflections and so the summation could be carried out. But very few other crystals could be dealt with.

There is, however, a way out. The problem can be reduced to two dimensions or even to one, although we obtain correspondingly less information. We have seen on p. 50 that it is possible for the indices of a reflection to have zero values, and that those reflections with one index equal to zero—the *hk*0s—are represented by one section of the reciprocal lattice (§ 4.5). If we use the structure factors (p. 58) derived from these reflections as coefficients in a Fourier series, the result gives us the projection of the electron density on the plane (001); if we use only the terms with two indices zero—say *h*00—we obtain the projection of the structure on to the *a* axis of the crystal.

7.3 *Introduction of the Fourier method*

The idea lay dormant until 1929. Then W. L. Bragg tried to see whether it was worth pursuing. He and West had made a measurement of the reflections 0*kl*, *h*0*l*, and *hk*0 from a crystal, diopside, $CaMg(SiO_3)_2$, and so he tried to see whether the Fourier summation would indeed give a picture of the electron distribution projected on to a plane with phases calculated from the atomic positions that he and West had just determined. An example is shown in fig. 7.6.

Of course, we now know that there could be no doubt. The theory is sound and the conclusion should be obvious. But when one is carrying out an operation for the first time, all sorts of worries enter one's mind. Can any theory really be trusted until it has been tested experimentally? Are the experimental data accurate enough? (Each point in the contour map of electron density depends upon *all* the measurements and slight errors in them may add together to make the whole

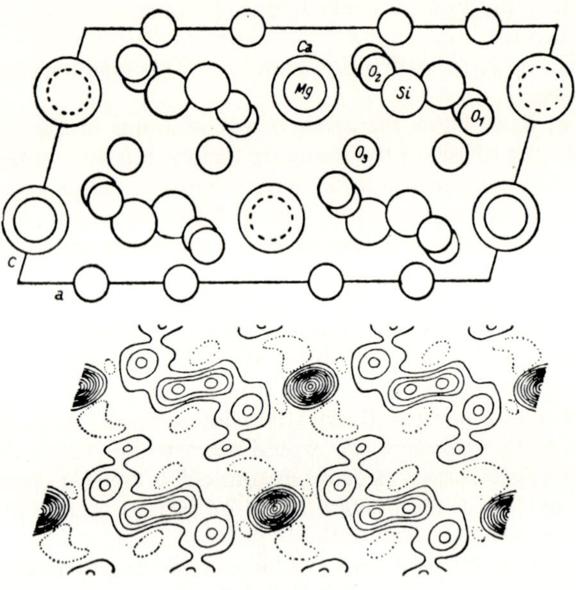

Fig. 7.6. Representation of the structure of diopside, $CaMg(SiO_3)_2$, projected on to the (010) face of the unit cell.

result meaningless.) Was it even possible to carry through the enormous computations involved without making so many mistakes that the result would not be recognizable? What Bragg showed was that these fears were groundless. Meaningful results *could* be obtained.

He seems to have carried out the work purely out of intellectual curiosity. This is no bad reason; many of the great scientific steps forward have been taken in this spirit, and not because they were felt to be stepping stones to further research. In his paper on his results, Bragg speculated whether the two-dimensional method could be used actually to *derive* structures. The phase problem (p. 99) had to be overcome, and he pointed out that, for one of the projections of diopside, it could be settled from symmetry considerations alone.

The Ca and Mg atoms lie on special positions in the unit cell, and in

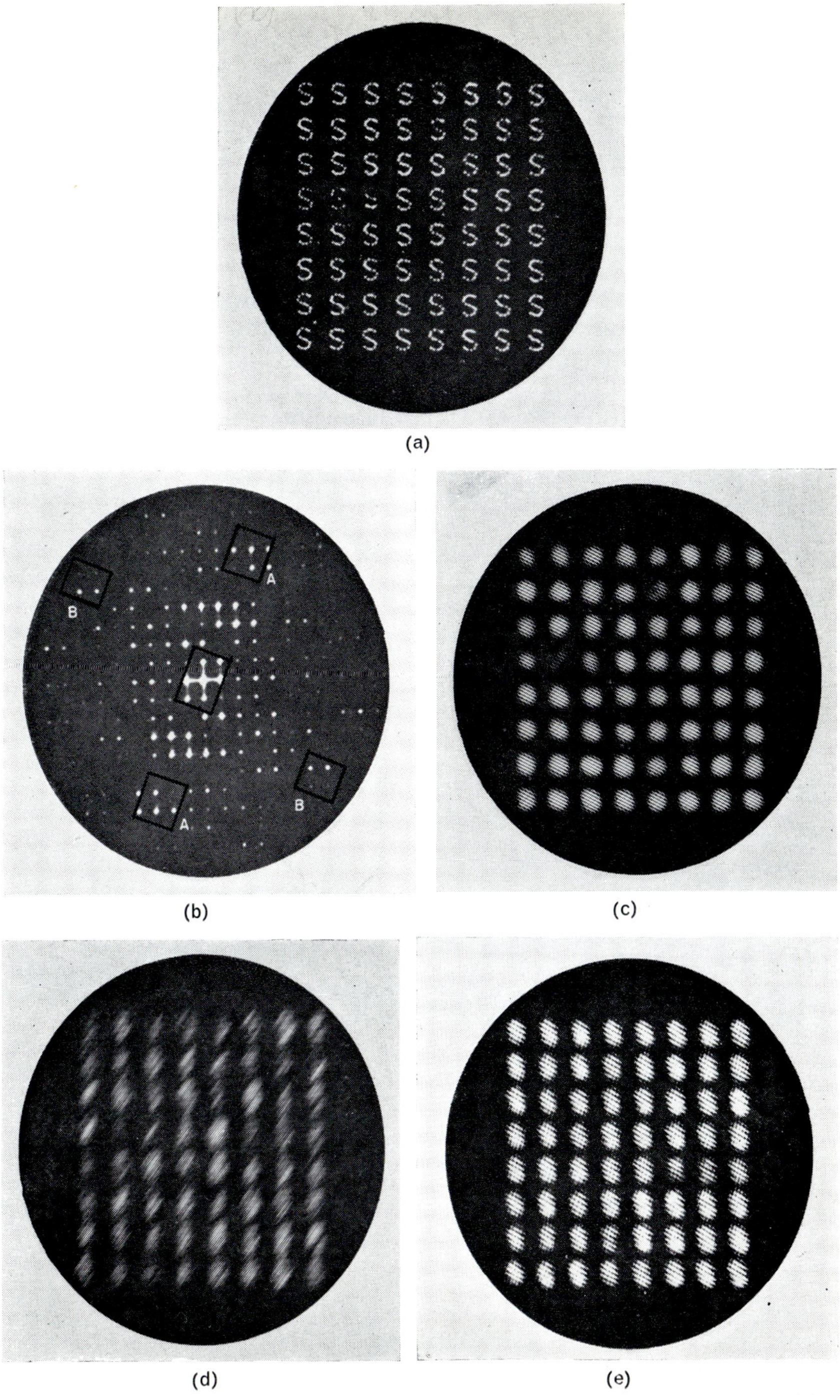

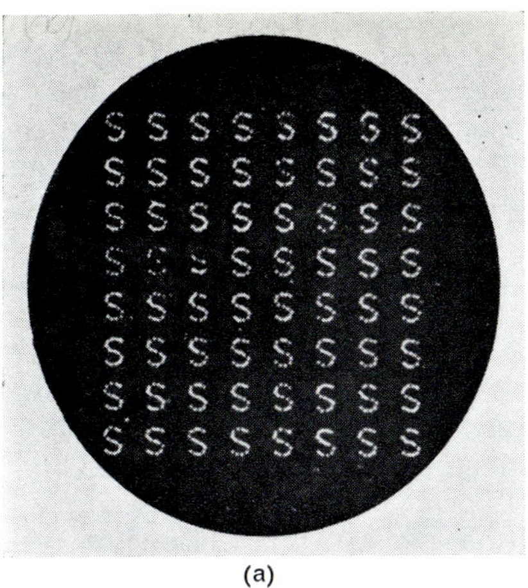

(a)

B A A B

(b) (c)

(d) (e)

Fig. 7.8. Formation of image of two-dimensional diffraction grating. (*a*) Diffraction grating with S motif; (*b*) diffraction pattern of (*a*); (*c*) fringes formed from parts of (*b*) labelled A; (*d*) fringes formed from parts of (*b*) labelled B; (*e*) fringes formed from parts AB and central part of (*b*), showing beginning of development of motif S.

Each pair of opposite orders gives a set of fringes, whose spacing depends upon the distance of separation of the orders, whose direction depends upon their direction of their separation, and whose intensity depends upon their intensities.

Each reflection in a pair of orders must have the same intensity (except in unusual circumstances which we shall discuss in the next chapter), but they will not have the same phases unless the unit pattern of the grating is centrosymmetrical. If the phases differ, the fringes produced will not have either a maximum or minimum at the origin. With these four variables of the fringes—spacing, direction in space, intensity and phase—the complete image can be built up (fig. 7.8).

In three dimensions the same principles apply, but, of course, in practice we cannot observe three-dimensional images. It is, however, extremely instructive always to think of an image as being formed of those very simple physical phenomena—Young's fringes.

CHAPTER 8
the Fourier explosion

8.1 *The first steps*

THE Fourier method, outlined by Bragg, had provided crystallographers with a tool that had two powerful attributes: first, it allowed more parameters to be handled at one time and, secondly, it was more objective than the methods of trial and error. One of the objections to trial-and-error methods is that a pattern of atoms had first to be assumed and then shown to be correct. How can we know that some other arrangements of atoms might not also give the same diffraction pattern? In fact, there is no proof that alternative arrangements are not possible, but, if a sensible structure gives good agreement, one assumes that any other set of atoms, even if it could be made to give acceptable agreement, would not make chemical or physical sense.

With the Fourier method, however, the same criticism would not apply. If one can, somehow or other, find the phases (§ 7.1) of the various structure factors, and then the Fourier synthesis shows peaks that one can recognize as atoms, with the right electron contents and the right distances apart, then one *knows* that the answer is right and that it is the only possible one.

Why, then, did not the crystallographic world immediately adopt the new idea and exploit its potentialities for solving structures on an increasing scale of complexity? There were several reasons. One is natural conservatism; while there were existing methods for solving problems of importance, why try new methods that might not be very successful and, in any case, involved a great deal of computation? After all, the method had been tried by Bragg only on a crystal of known structure, and everyone knows that a new method will always work on a solved problem!

Secondly, there were no standard methods of summing Fourier series. In one dimension it is possible to carry out the work by drawing curves and adding ordinates (p. 57). But even to sum a few terms in this way takes a long time. To carry out the summation in two dimensions, with 30–40 terms or even more, was rather forbidding.

Thirdly, of course, there was the phase problem, which meant that the method could be applied only in special circumstances.

One of the first attempts to solve an unknown structure was Cork's work on the alums in 1927, carried out in Manchester under the direction of W. L. Bragg. Here was a favourable problem: as we have seen, the positions of some of the atoms were known and therefore Fourier methods should help to find the positions of all the others. Cork, however, tried to reduce the problem to one dimension, by finding the projections of the structure on to the cell edge, the face diagonal and the cell diagonal; for these projections he needed to measure the *h*00 reflections (that is, all the orders from the (100) face), the *hh*0 reflections

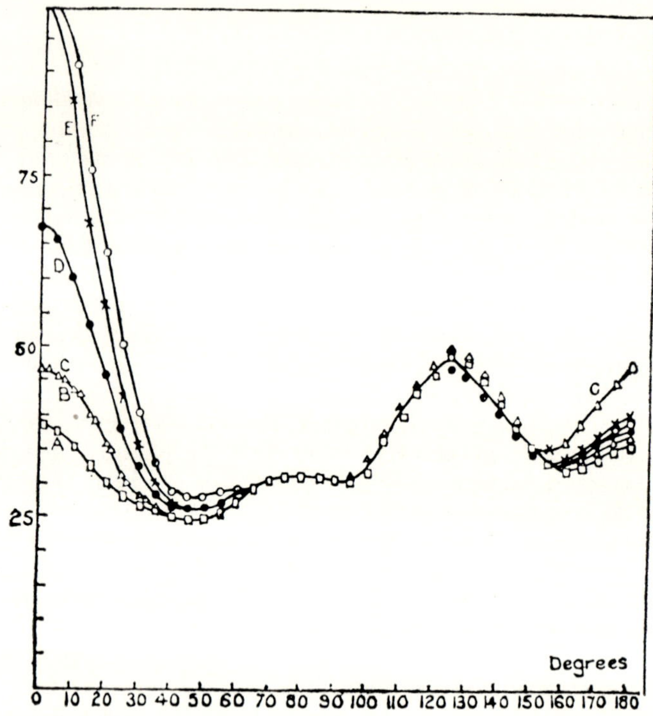

Fig. 8.1. Projections, on the diagonals of the unit cell, of the electron densities —in arbitrary units—in some alum structures. The curves are as follows: A—ammonium aluminium alum; B—potassium aluminium alum; C—potassium chromium alum; D—rubidium aluminium alum; E—thallium aluminium alum; F—caesium aluminium alum. 0° along the axis of abscissae represents the origin of the unit cell; the increasing electron density of the heavier monovalent atoms is clearly seen. 180° represents the centre of the unit cell; the curves here should all be the same, representing the electron density of aluminium, except C which represents chromium. In between there is a complicated distribution of sulphur and oxygen atoms, which is the same for all the alums.

(the orders from the (110) face), and the *hhh* reflections (the orders from the (111) face). Figure 8.1 shows the last of these three results; Cork measured the reflections from a large number of alums, and his curves show clearly the increase in electron content as the mass of the monovalent atom is increased, and as chromium replaces aluminium. But in between is a complicated curve representing the other atoms, and Cork could not interpret it.

He made the natural assumption that the intermediate peak represented sulphur, the heaviest of the other atoms. In fact, this was unlikely; since most of the other atoms are arranged in triplets around the three-fold axis (p. 37) each set would have more effect than a single S atom. Three O's contain 24 electrons; S contains only 16. Thus Cork's work did not take the problem much further and several years elapsed before the problem was finally solved.

A more productive research was that carried out by West, also in Manchester, in 1930, on the structure of potassium dihydrogen phosphate, KH_2PO_4. This was a highly symmetrical crystal, in which the potassium and phosphorus atoms were fixed by symmetry; since the scattering of the hydrogen atoms is negligible, only the oxygen atoms need to be found. West was able to sum a Fourier series, which showed clearly the positions of these atoms.

Although this was a great step forward, there is little doubt that the problem could have been solved by conventional methods, since it involved only three parameters—the x, y and z coordinates of one of the oxygen atoms. Trial-and-error methods could have coped with this degree of complexity without any difficulty. We must therefore seek other examples of really productive use of Fourier methods.

8.2 *Refinement of structures*

First, however, we must deal with a simple, but very important, use of Fourier methods—the determination of the atomic parameters with as great an accuracy as possible or what is called *refinement*. With trial-and-error methods, there was no means of knowing whether the best possible agreement between the calculated and observed diffraction pattern had been found; it was always possible that slight shifts here and there might improve the agreement still further. In fact, the refinement usually stopped when the research worker became bored, and decided that any extra gain in accuracy was not worth the trouble that it involved. This was not a satisfactory state of affairs.

The Fourier method gave a quick and complete solution to the problem. If the atomic positions were found accurately enough to give the right phases—particularly for a centrosymmetric structure for which they must be 0 or π—then a Fourier synthesis should give the best possible coordinates, and no further pushing about would be needed.

This method was used considerably in the 1930's, but later some doubts were raised about it. How did experimental errors affect the result? Were the missing reflections—those too small to be measured—important? What was the effect of introducing only a finite number of terms, when the Fourier series should, in principle, be infinite? These and other queries have reduced somewhat the importance of the

Fourier method of refinement, but they have certainly not removed it altogether.

An interesting side-line is what is called the *difference synthesis.* This is a Fourier synthesis with the differences between the calculated and observed structure factors as coefficients. Thus, if the observed value is 17 and the calculated 22, the difference necessary to produce the right value is -5; if the calculated value is -24, the structure factor is clearly negative and the correction is therefore $+7$.

The difference synthesis gives the difference between the electron densities in the true and the assumed structure. It gives no more information than is contained in the ordinary Fourier synthesis but, since it does not contain any full atomic peaks, it shows up the information

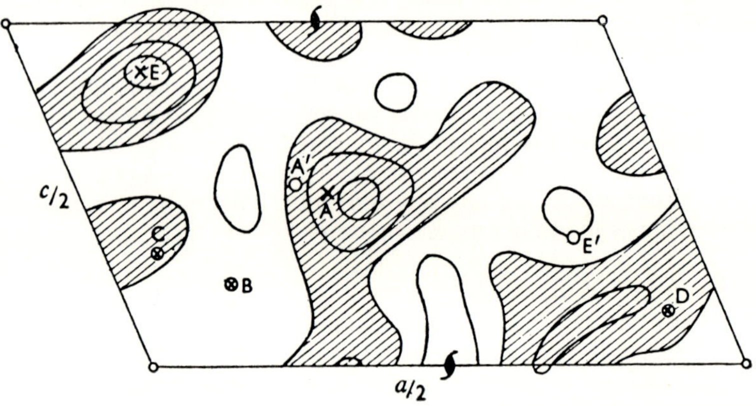

Fig. 8.2. Difference synthesis for an incorrect structure of durene, $C_{10}H_{14}$, with negative areas shaded. The atom at A should be at A′ and can be seen to lie on a steep slope. E, in a negative trough, is quite wrong; it should be at E′. B, C and D are correct.

more clearly. For example, if an atom is incorrectly placed, its assumed position will lie on a slope perpendicular to the direction of movement to the true position and the slope will give the amount of movement; if a wrong atom is assumed at a certain position—say, an oxygen atom instead of a carbon—then a negative region will be found there; and if some atoms, such as hydrogen, have been omitted from the calculations, the difference synthesis will show them clearly. These points are illustrated in fig. 8.2. There are other—perhaps better—methods of refining structures, but the difference synthesis is popular because of its simple physical meaning.

8.3 *The heavy-atom method*

Now let us return to the main business of Fourier synthesis—the determination of unknown structures. An opportunity arose to try out the method when an ordinary attempt on the structure of copper sulphate, $CuSO_4 . 5H_2O$, proved abortive. This was not surprising; it will be remembered that this was the crystal first used to diffract X-rays (p. 18) but it was not continued with because of its low symmetry. It is triclinic; that is, its unit cell is a general parallelepiped, and it has only a centre of symmetry. To specify all the atoms, we should need to know 33 parameters—three for each of the eleven atoms, excluding hydrogen. For the early 1930s this was indeed a formidable problem.

Perhaps it was not wise for two new research workers in Liverpool—Beevers and Lipson—to tackle such a problem. But sometimes one

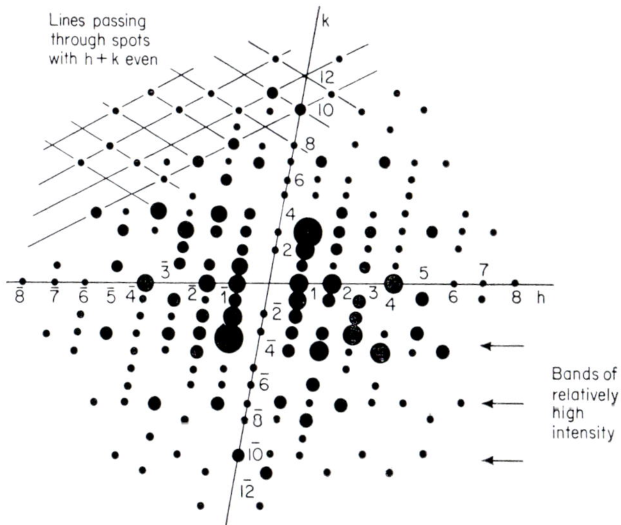

Fig. 8.3. Weighted reciprocal-lattice section for $CuSO_4 . 5H_2O$. At the top left-hand side lines are drawn through points with $h+k$ odd; most of the observed reflections lie on these lines. At the bottom horizontal bands of relatively strong intensity are shown.

learns a great deal by tackling problems that are too difficult, and necessity thrusts us into new approaches in order to solve them. In this respect, copper sulphate was certainly a good problem to choose. Although it took more than a year to obtain the solution, now it can be given as a simple, purely objective, exercise for students.

We shall consider only the projection of the structure on the plane (001) which requires measurements of the $hk0$ reflections; these were obtained on the ionization spectrometer in W. L. Bragg's laboratory. The values of F $(hk0)$ are shown on p. 110. It is always worth while seeing if such a table gives any clues and, in this case, two were fairly clear. They can be brought out by drawing a reciprocal-lattice section (p. 51) on which a black spot is drawn representing the size of the F

(fig. 8.3); this representation—called a *weighted reciprocal lattice*—is more graphic than mere numbers.

First, we can see that, in general, the reflections with $h+k$ odd are weak. There are some exceptions to this rule, but it shows up particularly well in the high orders. What does it mean? It means that

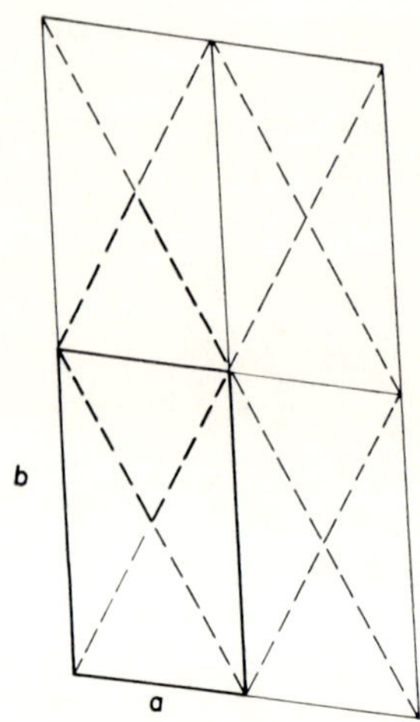

Fig. 8.4. Sections of the lattice of $CuSO_4 . 5H_2O$ showing four unit cells, one heavily outlined. The broken lines correspond to the pseudo-lattice given by the reciprocal lattice indicated at the top left-hand side of fig. 8.3.

k \ *h*	$\bar{8}$	$\bar{7}$	$\bar{6}$	$\bar{5}$	$\bar{4}$	$\bar{3}$	$\bar{2}$	$\bar{1}$	0	1	2	3	4	5	6	7	8
0	8	11	12	0	28	0	33	26	129	26	33	0	28	0	12	11	8
1		11	0	20	0	21	18	30	0	18	10	19	10	0	0	0	
2		0	0	0	0	21	0	17	12	32	17	7	10	0	9	0	
3		0	0	18	7	26	20	23	11	50	17	18	7	16	6	8	
4		0	18	13	26	8	27	17	0	0	17	10	8	0	10		
5		0	0	0	0	12	0	0	14	11	15	0	10	0	0		
6		0	0	0	7	0	16	0	12	14	15	0	9	0	0		
7		7	0	13	9	19	15	21	15	18	0	19	7	15			
8		0	10	0	0	0	16	8	12	0	11	0	0	0			
9			0	0	0	0	11	0	0	0	0	7	0				
10			8	0	12	0	14	7	16	0	10	0	12				
11				13	0	12	0	17	0	8	0	7					
12				0	0	0	0	0	0	0	0						
13					0	0	0	11	0	7							

there is some regularity in the structure that makes it approximate to a smaller unit cell, such as that shown in fig. 8.4. The simplest explanation is that the heaviest atoms in the structure occupy centres of symmetry as shown in this figure. It is not the only explanation, but the others led to blind alleys and we shall not consider them here.

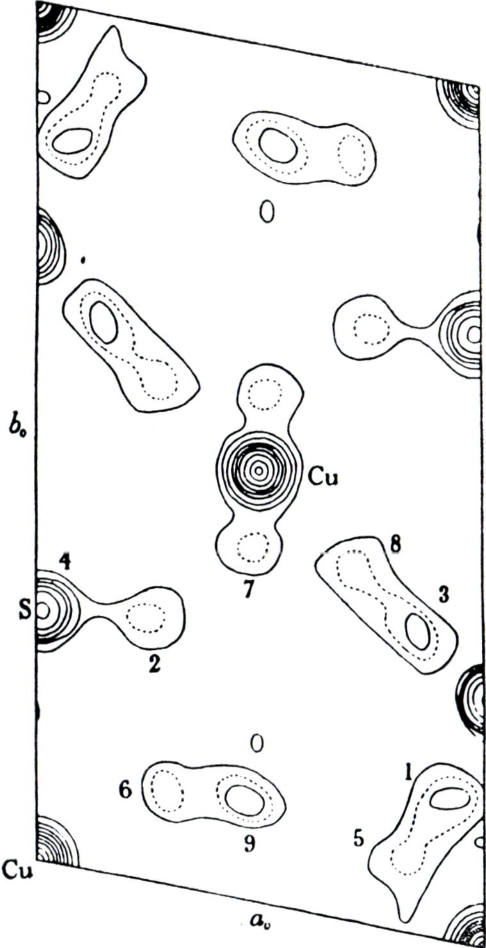

Fig. 8.5. Projection of the structure of $CuSO_4 . 5H_2O$ on the (001) face of the unit cell. The Cu and S atoms are marked, and the O atoms are indicated by the numbers 1 to 9, atom 4 being on top of S. There are two smaller peaks that are not significant.

This is a great step forward. We have fixed the two heaviest atoms, and the number of parameters left is now 30. What can we see next? Well, we notice that, of the reflections with $h+k$ even, there is a tendency for reflections with the same k to be similar. Thus, nearly all those with $k = 2$ are small and those with $k = 3$ and 4 are large (fig. 8.3). This evidence is sufficient to fix the positions of the next heaviest atoms, the sulphurs, at $x = 0{\cdot}00$, $y = 0{\cdot}29$ (note that $0{\cdot}29 = 1/3\frac{1}{2}$).

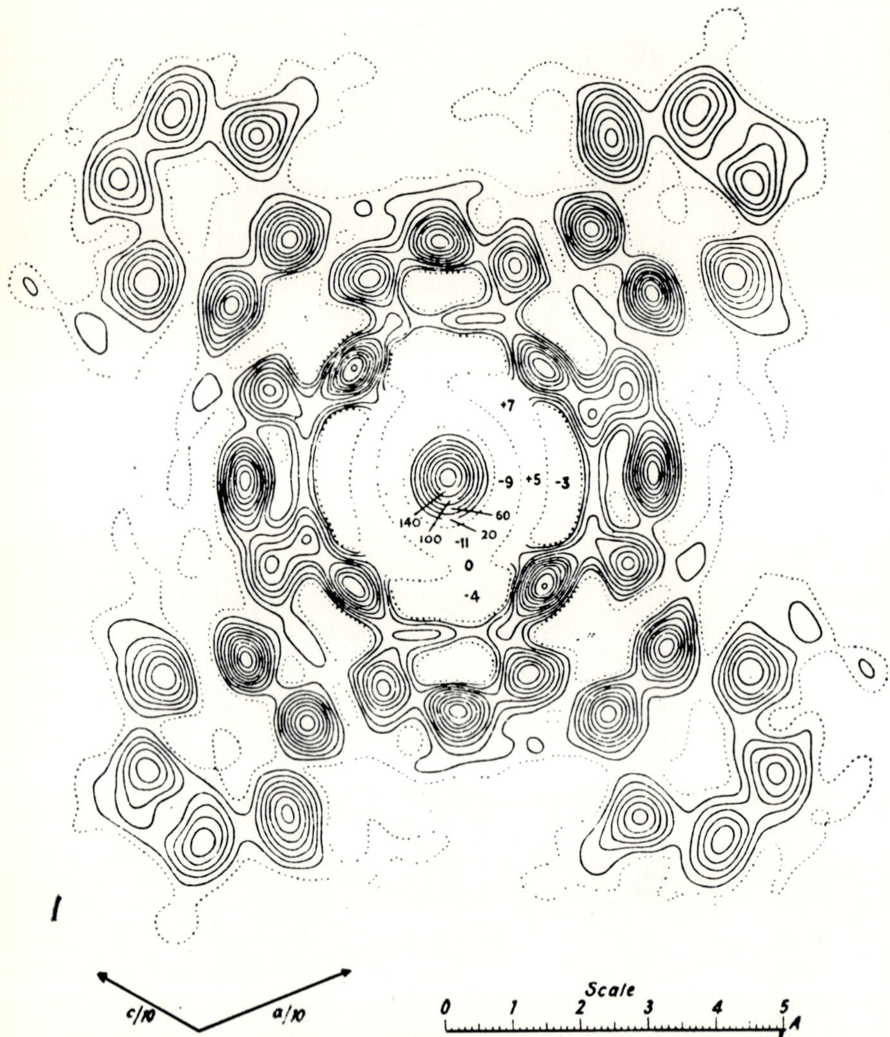

Fig. 8.6. The molecule of phthalocyanine, with a Pt atom at its centre.

This is as far as Beevers and Lipson could go. They knew that the SO_4 group should have the form of a regular tetrahedron, but they had no idea of how the unit cell should accommodate five water molecules.

How can one accommodate an odd number of atoms around a central atom?

Then the idea of using Fourier methods arose. As we have pointed out, they had not been used for a crystal of this complexity, and efficient methods for summing the series were not known. But since there was no other way of finding the structure, Beevers and Lipson decided to proceed with the work using the Cu and S atoms to decide the phases. Preliminary tests indicated that the computations would take nine months—a prohibitively long time; but by systematizing the work, as we shall explain in § 8.8, they finished it in one month.

The computations were completely successful. The result, shown in fig. 8.5, is clearly explained in terms of the right number of atoms. Bragg's idea of using the heavy atoms to fix the phases of the structure factors (p. 58) had been shown to work, and the way was now open for tackling other problems.

The method—called the *heavy-atom method*—is now quite standard. If the structure of an important chemical compound is required, a heavy atom such as bromine is attached and its position found; this step is usually easy because, as with copper sulphate, the heavy atom impresses information about its position on the diffraction pattern. The phases are calculated, and the Fourier synthesis computed. The atoms should appear clearly, and the structure can then be refined by standard methods.

Physically, the method is equivalent to focusing a microscope upon a specimen that contains no clear detail; one cannot be quite sure where the correct plane of focus is. But if we put a mark on the specimen, we can focus on that, and the rest of the pattern should be in focus with it.

One of the most striking results obtained with the method was the structure of phthalocyanine, by Robertson and Woodward, working in W. H. Bragg's laboratory in London in 1936. This is a plane molecule, in which a metal atom can be placed at the centre. In the crystals, the metal atom occupied the origin of the unit cell, and this gave a positive contribution to all the structure factors (p. 58). For platinum phthalocyanine, it could be assumed that all the structure factors were positive, and a Fourier synthesis then gave a complete representation of the molecule (fig. 8.6) without any further complications.

8.4 *Patterson's synthesis—interatomic vectors*

Soon after the introduction of Fourier methods, a new procedure was suggested that looked, at first, as though it was the answer to the phase problem. The American crystallographer, Patterson, in 1935 showed theoretically that, if one used the squares of the structure amplitudes (p. 58) as coefficients in a Fourier series, the resulting synthesis gave direct information about the structure; since these coefficients were all positive, there was no phase problem and therefore no doubt at all about the result.

Patterson—who called the result an F^2 synthesis although everyone else called it a Patterson synthesis—showed that the peaks represented interatomic distances; if there were a peak at (x,y,z) in the Patterson synthesis it must mean that there were atoms whose coordinates differed by these values. If the peak were strong, it meant either that the two atoms were heavy, or that there were several atoms related in this way.

This information was so direct that it seemed that the problem of deriving the atomic coordinates themselves must be quite simple. Patterson supported this claim by using the published data for copper sulphate (p. 110); he showed that it was possible to derive the positions of all the atoms directly, with only some slight ambiguity.

But we now know that the method is limited. Its weakness is that the number of interatomic vectors is so large; if there are N atoms in the unit cell, there are $N(N-1)$ vectors, since each of the N atoms has $N-1$ companions. When N becomes large—even into double figures—$N(N-1)$ becomes unmanageable. The reason why the application to copper sulphate was successful was the same as that which made the initial method work: the heavy copper atoms lay in special positions (§ 6.6). The Cu–S and Cu–O vectors were easily found, and the O–O vectors formed a generally featureless background that did not upset the information already derived.

The method is much more powerful in three dimensions than in two. Suppose that we know that no two atoms are closer than 1·5 Å. Then in three dimensions all the nearest-neighbour vectors must be represented by peaks on a sphere with radius 1·5 Å around the origin; in two dimensions, peaks can be at any distances because atoms may overlap in projection.

Although the Patterson method, therefore, was not a complete answer to the problem, it has settled down to a useful place in the subject and some structures have been directly derived by means of it. It is often regarded as an obvious first step in a structure determination; the result is easy to derive, it is objective, and it may possibly provide some helpful information. But its most important contribution is the determination of the heavy atoms for use in the method described in the last section. In this way it has played a great part in many of the researches to be described in the following sections.

8.5 *Isomorphous replacement*

As soon as it was evident that the Fourier method was practicable, other approaches were sought. Beevers and Lipson decided to try to complete the structure of the alums (p. 106) and this time they decided to see whether accurate measurements were really necessary. It had taken many weeks of work to measure the $hk0$ intensities for copper sulphate, and if eye-estimated values were good enough a great deal of time could he saved. Of course, for really precise work, accurate measurements were certainly necessary, but for establishing the rough atomic positions they might not be needed.

Other difficulties occurred, however, before this point could be settled. As we have seen on p. 91, the potassium and aluminium atoms have the NaCl arrangement which is face-centred; that is, they contribute only to those reflections with *h*, *k* and *l* all even or all odd: but the complete structure is not face-centred, and therefore the potassium and aluminium atoms do not contribute at all to the reflections with mixed

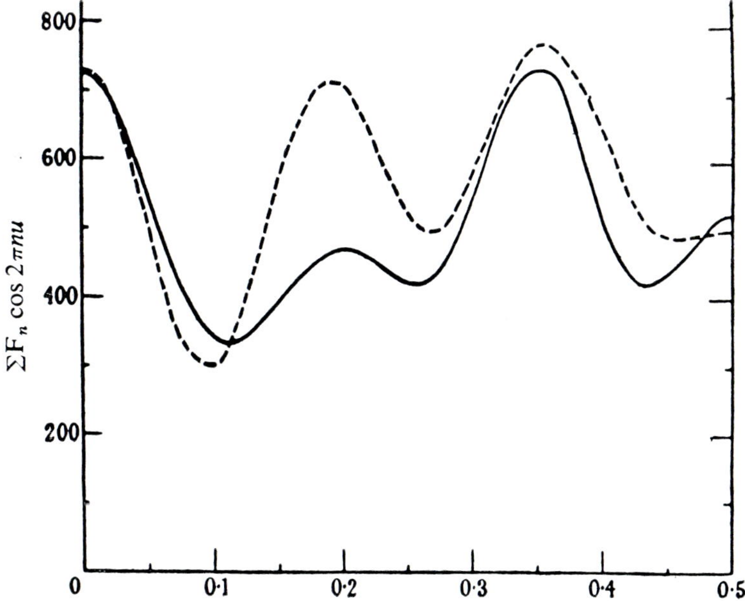

Fig. 8.7. Comparison of the electron densities in arbitrary units projected on the cube diagonal in $KAl(SO_4)_2 12H_2O$ (full line) and $KAl(SeO_4)_2 12H_2O$ (dotted line) (compare fig. 8.1). It can be seen that the large peak in fig. 8.1 does not correspond with the S atoms, as Cork supposed. (The abscissae represent distances along half the diagonal, and are represented as fractions, not degrees as in fig. 8.1.)

indices—even and odd. Thus the heavy atoms cannot give information about the phases of these reflections. Of the *hk*0 reflections, which are needed to produce a projection of the electron density upon the (001) plane, only half can be dealt with; one might expect to obtain a recognizable picture if some reflections are omitted, but not if they amount to so many.

With copper sulphate, the problem was solved by finding the position of the sulphur atom, but for the alums we have seen that the position of this atom could not be found (p. 107). Beevers and Lipson therefore decided to make use of the selenate, in which selenium replaces sulphur.

Using Cork's method with the *hhh* reflections from potassium aluminium selenate, they found a new peak (fig. 8.7) on the cube diagonal and were thus able to identify the position of the sulphur atom.

The phases of most of the structure factors were now obtained—enough to give a recognizable picture which could be refined (§ 8.2). The method is now called the *isomorphous-replacement method.* It supplements the heavy-atom method and the two together have been responsible for a large proportion of the crystal structures that have been determined.

One of the most striking examples is given by the well-known chemical compound, strychnine. The exact form of the molecule had evaded the chemists for many years and the crystallographers decided to see if they could help. In 1948, at Utrecht in Holland, Bokhoven, Schoone and Bijvoet determined the structure by means of the sulphate and the selenate, and in 1950 Robertson—not the same Robertson who worked on phthalocyanine—and Beevers, in Edinburgh, confirmed the result with the hydrobromide. By this time, the chemists had also arrived at the solution by orthodox chemical methods; nevertheless the research marked the occasion when the chemists had to accept X-ray crystallography as a serious contributor to the art of finding the structure of an unknown chemical molecule. Up to this time, X-ray methods had merely confirmed what the chemists knew already—except for NaCl (p. 82)!

8.6 *Penicillin*

Now comes one of the great episodes in the subject—the determination of the structure of penicillin. Like so many important researches, it does not fit neatly into any one category, but it nevertheless illustrates clearly the spirit in which this sort of research is carried out.

The importance of the problem was clear. Penicillin was an extremely important compound during the Second World War, but its method of manufacture—through the growth of moulds—was rather odd; it was thought that, if its structure could be determined, chemical methods might be found for manufacturing it in a much more straightforward way. X-ray crystallography was therefore one of the methods that had to be tried.

A search thus began for a crystalline compound containing a heavy, and if possible replaceable, atom; the compound had to be well crystallized and the unit cell had to be reasonably small so that it did not contain too many molecules. Finally the sodium, potassium and rubidium salts of benzyl penicillin were chosen. But, as usual, there were unexpected difficulties; the sodium compound was not isomorphous with the others, being monoclinic instead of orthorhombic. And when the two orthorhombic compounds were examined, it was found that the heavy atoms had one coordinate that caused them to have no contributions to a large proportion of the reflections. It was the alum story once again.

This work now splits into two parts—the orthorhombic part under Crowfoot (Mrs. Hodgkin) at Oxford and the monoclinic part under Bunn working for I.C.I. at Northwich. Quite different methods were used by each: the former used the isomorphous-replacement method as far as it would go, and then tried to proceed by guesswork supplemented by chemical information; the latter used optical analogues that will be mentioned in Chapter 9. The characteristic of the great crystallographers—amongst whom Mrs. Hodgkin is outstanding—is that they can see intuitively which method is likely to be of value in any particular problem; they do not feel bound to use a specific approach for all their problems.

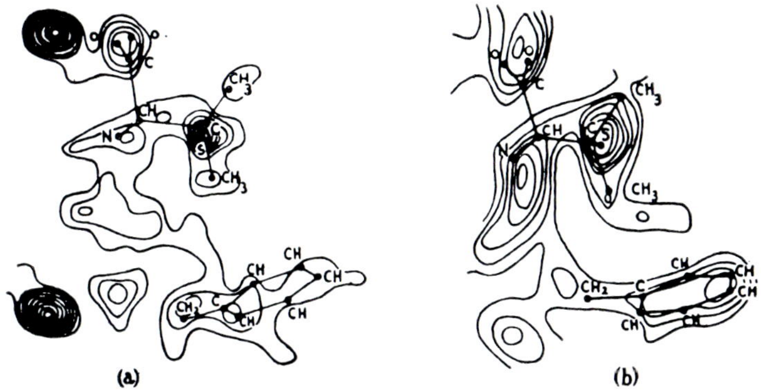

Fig. 8.8. Comparison of detail in early Fourier syntheses for rubidium benzylpenicillin and sodium benzylpenicillin.

For penicillin, this unorthodoxy paid off; although neither side could achieve success, when they compared results they found certain features in common (fig. 8.8). Adopting these features and neglecting the others, they were able to construct a model that gave an acceptable agreement between calculated and observed intensities. Thus although the sodium salt was not isomorphous with the others, it had yielded results which, when compared with the others, had been just as good as if it had been isomorphous. In fact, the completely different approach had, if anything, made the comparison of the common features of the two results even more convincing.

Unfortunately, it cannot truthfully be claimed that the ultimate object of the research was achieved. Certainly, we now understand the chemical structure of the molecule of penicillin, but this does not help us to manufacture it by a chemical process. It is still produced from moulds!

8.7 *Direct methods*

These various methods, successful though they were, did not satisfy some people. They would have preferred to work in what is supposed by the layman to be the typically scientific approach—to put one's data into a set of formulae and to produce an answer directly without any assumptions. But most crystallographers believed that this was impossible; because one could not observe the phases of the X-ray waves, direct methods were bound to be impossible.

This attitude was changed abruptly in the most effective way possible: in 1948 Harker and Kasper, at Schenectady in U.S.A., worked out a new structure by a completely mathematical method that involved no assumptions at all. The compound was decaborane, $B_{10}H_{14}$. It contained no heavy atom, and packing principles applied in an orthodox way gave no help. As the workers said afterwards: 'The structure would not come out by ordinary methods, so we *had* to find a new approach.'

We can do no more here than give a brief idea of how the method operated. Let us suppose that we try to work out a structure by taking phases at random, producing a Fourier synthesis, and seeing if anything like a recognizable result emerges. Let us repeat this operation with many different combinations of phases. (This is not a sensible approach, but in principle it is possible). Most of the answers will be quite meaningless; they will not contain peaks representing atoms, and in some places there will be troughs going well below zero electron density. Can we impose any conditions on the phases that will prevent this latter occurrence?

Harker and Kasper found that, for certain combinations of phases of some of the strong reflections, negative electron densities were bound to occur at certain points in the unit cell, whatever the phases of the rest of the reflections. (This interpretation was not appreciated by them at the time; it was introduced later.) Thus these combinations could be dismissed. By gradually introducing more strong reflections, they were able to fix the phases of most of them, and they then summed the Fourier series. This gave a recognizable picture of the structure—a picture quite different from what they had expected and which caused a revolution in the understanding of the chemical behaviour of boron.

The method was not of very general application; it could be applied only to crystals that gave some particularly strong intensities and these occurred only if there were relatively few atoms in the unit cell. But, once the barrier had been pierced, other people helped to open it further. More elaborate theories were introduced and methods involving digital computers were devised. Now structures much more complicated than $B_{10}H_{14}$ can be successfully tackled by these methods, and finality in their approach has not yet been achieved. It will be interesting to see, over the next few years, how far these methods can go.

8.8 *Anomalous scattering*

In 1949 another experimental approach was introduced—one that had been known about since the beginning of X-ray diffraction but which was thought to be too insensitive to be of any value. It makes use of the different relative scattering factors of atoms when they scatter different radiations.

We have stated on p. 55 that the scattering factors of atoms are functions of $(\sin \theta)/\lambda$ and thus that the relative values for different atoms should not change for different wavelengths so long as $(\sin \theta)/\lambda$ is constant. There are, however, exceptions to this rule; these occur when the frequency of the radiation is near to that necessary to dislodge an electron from an atom. The scattering is then said to be *anomalous*, and even if $(\sin \theta)/\lambda$ is the same for two different values of λ, the scattering factor of the atom will be different.

The physical basis of this effect can be seen by performing a simple experiment—possibly the most rewarding experiment in the whole of physics if we relate the amount of information obtained to the effort

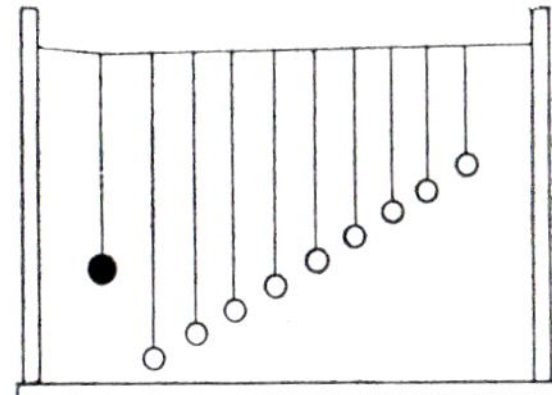

Fig. 8.9. Experiment for illustrating resonance. Ten pendulums are shown, supported from a light string. The black blob represents a heavy lead ball; the others are made from wood.

involved in carrying it out. What we wish to do is to see how one oscillator can affect others of different frequencies. Let us provide what we can call a master oscillator by making a simple pendulum about 1 m long with a heavy lead bob. Attach this to a horizontal string, tightly supported at both ends. To the same string attach other pendulums of different lengths (fig. 8.9), with less heavy bobs; one should be of about the same length as the master pendulum.

Now set the master pendulum in motion. It will oscillate slowly and since it is connected—the scientific word is 'coupled'—to the other oscillators, they will also be set in motion. However, the one with the same length as the master pendulum will have a far greater amplitude than any of the others. This is the well-known phenomenon of *resonance*.

But let us look closely at the other pendulums. When all their motions have settled down, they can be seen to be vibrating with the same frequency as the master pendulum. But those that are of longer length than the master are vibrating in the same phase—that is, they are displaced in the same way as the master at every instant of time—and those that are shorter are vibrating out of phase. If we had a very large number of driven oscillators we could trace the way in which the oscillation changes from being in phase to being out of phase; the theory is too complicated to be considered here.

The application to X-ray diffraction lies in considering the master pendulum to be the incident X-ray beam, and the driven oscillators to be the electrons in the various energy levels of the diffracting atoms. Since these electrons oscillate with the same frequency as the incident beam, the scattering is in phase, as we have assumed in the theory discussed so far. But in the region near to resonance odd changes in phase can take place; all the electrons do not now scatter in phase, and the theory that we have outlined is no longer adequate.

The theory works satisfactorily because most of the natural frequencies in atoms are less than those of the incident X-rays normally used. The K electrons—the most firmly bound in the atom—may have energies with frequencies near to those of the X-rays, and these electrons can therefore be affected in the way that we have described. The effect is not large, however, because there are only two K electrons in each atom, and to a first approximation the anomalous scattering can be ignored.

In 1949, however, Bijvoet showed that it was appreciable, and that, by choosing two radiations with appropriate wavelengths, one could have two different scattering factors for the same atom. For example, Peerdeman, van Bommel and Bijvoet applied the method to the crystal sodium rubidium tartrate, using ZrKα radiation; the wavelength of this radiation is about 0·788 Å, and the wavelength associated with the K shell of rubidium is equal to 0·815 Å. They found some changes in relative intensity, compared with those given by CuKα radiation, of up to 10%, and they were able to ascribe these changes to the known structure of the compound. The effect has since been used to determine unknown structures in which the heavy atom by itself was not sufficient to determine an adequate number of phases and isomorphous compounds could not be obtained.

But the method has another, very important, property; it enables us to find the absolute configuration of a molecule. Thus, if we have a carbon atom with four different groups attached to it tetrahedrally (fig. 8.10) we do not know whether it has the form shown in (*a*), or whether it has the mirror image (*b*). Bijvoet's method enables us to make this distinction; in other words, it enables us to distinguish between right-handed and left-handed optically active compounds.

8.9 *Methods of summing Fourier series*

Summing a one-dimensional Fourier series is not difficult, although the work can be very tedious, particularly if the number of terms is large. Two-dimensional series are a different matter; usually a large number of terms is involved and since the series has to be summed over a two-dimensional surface the work can be quite formidable. As we have seen on p. 105, the difficulty of summing series was one reason why there was some reluctance to try out the Fourier method in the 1930's. The present section is not intended to give a working picture of how the process was first started; it is designed merely to point out the problems involved and to show in a general way how they were overcome.

When Bragg summed the series for diopside in 1928, he did not describe how he had carried out the work, and so left no guidance for other people. When Beevers and Lipson started on copper sulphate (p. 113) they had little idea how to begin. It was clear that adding the terms, one at a time, at grid points covering the whole area of projection,

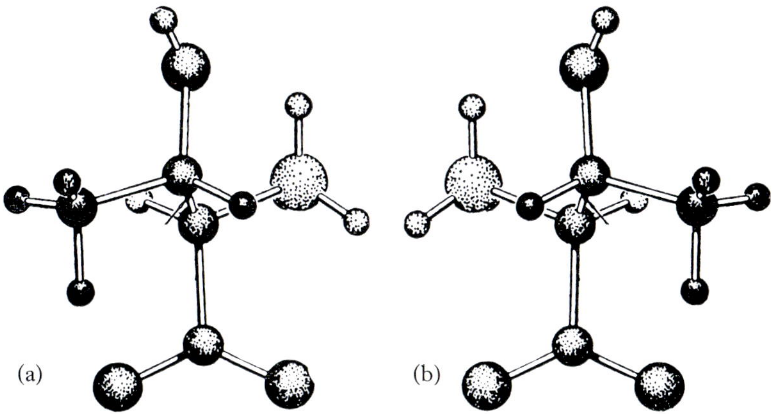

Fig. 8.10. Two molecules related by a mirror plane—that is, as the left hand is to the right hand.

was impracticable; as stated on p. 113, a trial showed that it would take nine months to carry out the complete operation this way!

They therefore devised a method that gave the summation along a line. It involved long strips of paper with numbers on; these numbers were read off successively for the first wave, alternately for the second, every third for the third, and so on. One worker read the numbers and the other multiplied by the amplitude on a slide rule. The numbers were entered in tables to two significant figures and then added mentally. The whole operation took about a month.

This was still too long. But while the process was in operation, new ideas were being hatched. The one that proved decisive was to make use of the equation:

$$\cos 2\pi(hx+ky) = \cos 2\pi hx \cos 2\pi ky - \sin 2\pi hx \sin 2\pi ky.$$

It may seem odd to try to simplify an operation by doubling the number of terms, but the expansion opened up possibilities of systematization. The expression $F(hk0) \cos 2\pi(hx+ky)$ could be written as:

$$\{F(hk0) \cos 2\pi hx\} \cos 2\pi ky \\ -\{F(hk0) \sin 2\pi hx\} \sin 2\pi ky,$$

the terms in brackets being considered as amplitudes of ordinary sine and cosine curves.

In addition, instead of entering the numbers in tables, they were written on strips of card and filed, so that they could be used again if required. Gradually a good stock of cards was built up, and other people made requests to use them. With their aid, a synthesis such as that for copper sulphate could be carried out in 3–4 days.

Then sets of cards—Fourier strips—were printed, so that they could be made available to other laboratories, and they have proved to be very popular. In spite of other, rather more accurate but more complicated, devices, they have been the most used method in the subject. One worker enthusiastically called them 'the first step in the scientific approach to crystal structures'.

Now, of course, they are outdated. With large unit cells and three-dimensional work, the operation of using the Fourier strips is too arduous. Yet they still perform a useful function—in introducing new research students to the idea of Fourier synthesis. There is still a small demand for them from laboratories that do not want to cut off their students completely from their roots, and who therefore demand that they shall perform at least one Fourier synthesis for themselves.

Everyone now uses the digital computer for serious work. It can perform in a few seconds work that would take weeks by the Fourier strips. One must not, however, think that the results are *available* in a few seconds; a few days usually elapse before the computations are returned, and even then they are not always correct. Finality is not with us yet.

8.10 *Ultimate achievements*

With all these methods available, how do we set about the determination of the structure of a crystal? There is still no general method that can be used to give an answer and the way that is chosen will depend upon many factors—the nature of the problem, the particular abilities of the investigator, the resources of money and assistance that are available to him. But it is now possible to indicate a general plan which, if it can be followed, is almost certain to work.

First of all, if one is interested in a particular material, one must try to find a crystalline compound with a heavy atom in it, unless the compound contains one already. What constitutes a heavy atom depends upon the number of light atoms present: with 20 carbon atoms, an atom of atomic number around 30 would do; with 100 carbon atoms, an atomic number of 60 would be necessary. If the atom is replaceable by one of another atomic number, so much the better.

For crystals that are not centrosymmetric, the problems are tougher, and isomorphous replacement becomes necessary. The more complicated the molecule, the less likely is it that the crystals will be centrosymmetric, since nature does not usually make the pairs of mirror-image molecules that are required for producing centrosymmetric crystals. With accurate data from these crystals, obtained by means of a diffractometer, it should normally be possible to work out the structure by Fourier methods.

There can be no doubts of the success that these methods have had. By means of them, Hodgkin and her co-workers at Oxford derived the structures of vitamin B_{12} (fig. 8.11) and of insulin (fig. 8.12), molecules containing over 100 atoms. The former defied the rules that we have just stated for the work was based upon a single 'heavy' atom of cobalt.

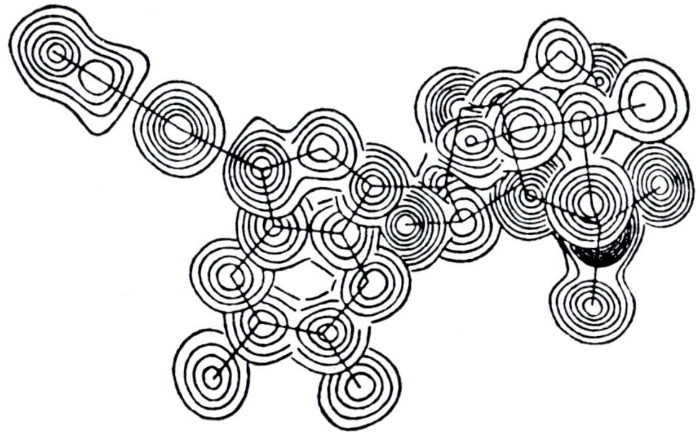

Fig. 8.11. One of the Fourier syntheses used in deriving the structure of vitamin B_{12}.

For most people this would have been far too small, but the rules do not allow for exceptional people! It was with this work in mind that we said that the approach depends upon the abilities of the investigator.

Then—still more complicated—there is the work on the proteins, by Perutz and Kendrew at Cambridge. Here the molecule was larger still, with about 5000 atoms. For these structures, heavy atoms—mercury—were deliberately introduced, held by forces that are not finally understood. Anomalous scattering (§ 8·8) was also used, and the structures were derived successfully. These structures are the most complicated that have yet been solved although not in as complete detail as vitamin B_{12}, and they have added considerably to the knowledge of the pattern of living matter. An immense range of complexity has been covered in the sixty years since the structure of NaCl was determined.

It is possible that these structures represent the ultimate of what can be accomplished. It may seem rash to make such a statement in the light of past rates of progress, but we make it deliberately. It may not be the limit of what could be done, but it must be near to the limit of what the human mind can adequately absorb. It is no use working out crystal structures unless they teach us something; usually they can be understood only by means of models. Adequate models of the proteins are

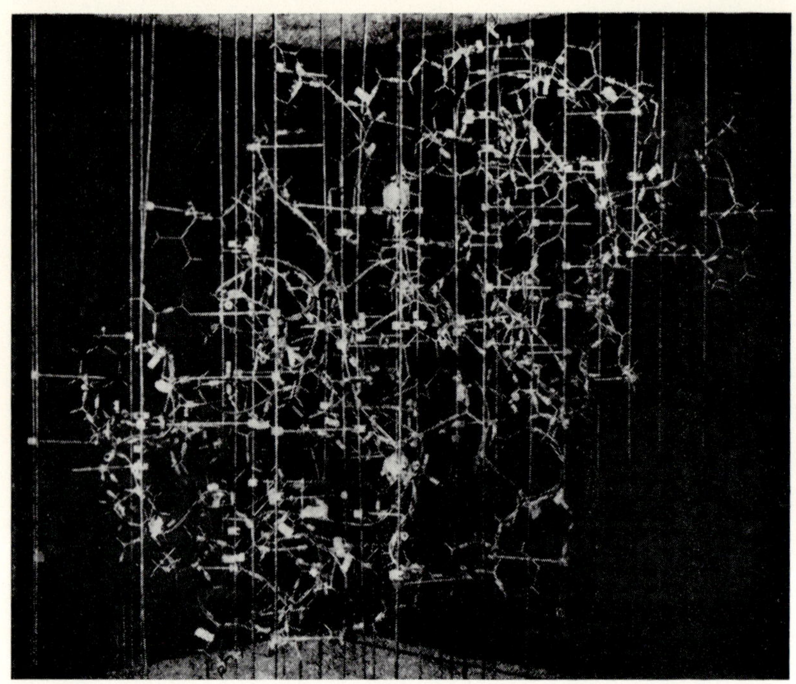

Fig. 8.12. The structure of insulin derived by X-ray methods.

difficult to make; another order of magnitude would mean that they were quite out of reach to the ordinary scientist with ordinary resources.

But there is no need for gloom. There is still plenty to do in the gap between the proteins and more ordinary crystals. Crystal-structure work is not likely to dry up.

CHAPTER 9

information derived from crystal structures

9.1 *General survey*

In the early days of crystal-structure work, problems were chosen more or less at random. Deriving a structure gave a glow of satisfaction quite independent of the information that the result might give. But this attitude could not last; it was gradually realized that it was no use compiling results without any aim. The new methods were capable of helping forward other branches of science, but they would do so only if the problems to which they were applied were chosen with discrimination.

Fortunately, there were some people who soon realized this and were able to direct the work into its most fruitful channels. They did not necessarily make contributions themselves, but they could see clearly what ought to be done. They could recognize what information could be obtained from crystal structures and they could indicate compounds likely to provide this information. Such people are needed. They are the ones who can take an overall view, without which science would become a conglomeration of undigested facts and results.

We now know that X-ray crystallography has made enormous contributions to the whole of the rest of science. In this chapter we shall try to show what these contributions have been, particularly in the fundamental understanding of the forces that hold atoms together in crystalline solids.

9.2 *Nature of interatomic forces*

It will be remembered that the first result of crystal-structure determination—the atomic arrangement in NaCl (p. 82)—revised completely the chemists' ideas concerning molecules. They had thought it so natural to ascribe the composition NaCl to the existence of molecules that no other possibility was seriously considered. When the Braggs showed that the crystal was a sort of 'three-dimensional chess board', as one chemist rather contemptuously called it, there was widespread disbelief. But the evidence was completely convincing and the result had to be accepted.

What did this mean? Each atom of sodium was in contact with six atoms of chlorine and vice versa, but no atom was in contact with one of its own kind. The simplest analogy was that of oppositely charged bodies; unlike charges attract and like charges repel. Are the two atoms oppositely charged?

Now, charges on atoms were known. In order to explain the phenomena of electrolysis, Faraday had had to postulate the existence of such particles in solution. It was believed that, when salt dissolved in water, the molecules dissociated into charged ions—positive sodium ions and negative chlorine ions. How much simpler it was to accept that the ions existed already in the crystal; they were held together by electrical forces, which decreased in solution because of the high dielectric constant of water.

This idea is now completely accepted. Molecules of inorganic salts do not exist. The type of force holding ions together is known as the *electrovalent* or *ionic* force.

It will be seen that this result had emerged without any quantitative information; the relative dispositions of the atoms were enough to introduce quite new ideas. From this simple basis, however, more complicated possibilities arose. Na^+ and Cl^- are quite simple ions; what happens with ions, such as SO_4^{2-} which are composed of numbers of atoms and thus are larger and have more awkward shapes? The ways in which crystals cope with these problems will be discussed when we deal with the ionic bond in more detail.

Ionic bonds can occur only when there are two sorts of atoms. Entirely different sorts of forces must apply when the atoms in a crystal are all similar to each other. For example, what forces hold the atoms together in a crystal of an element? Let us take sodium as an example.

The structure of sodium is body-centred cubic, with atoms at the corners of the cubic unit cell and at the centre. Thus each atom is surrounded by eight exactly similar atoms. The nature of the forces holding them together cannot be obtained by deduction, as it could be —more or less—for NaCl; we must bring in some of our general knowledge about metals.

A very large proportion—about three-quarters—of all the elements are metallic; they are ductile, they conduct electricity, they are opaque and they reflect light in a characteristic way which we all know but which we cannot define. In the periodic table metals occupy the left-hand side, gradually increasing their penetration to the right as their atomic numbers increase. Thus, while of the first ten elements only two—Li and Be—can be properly said to be metallic, of the first 40 the number has increased to 23.

The structures that these atoms possess are characterized by a small number of electrons in the outer shell. Sodium, which we have taken as a prototype, has one electron in the K shell, eight in the L shell and one in the M shell. It is therefore natural to assume that the metallic properties of sodium are produced by this loosely bound electron; it can move about from atom to atom, thus accounting for electrical conductivity, and it can respond to electromagnetic waves, thus accounting for the reflection of light.

It is this so-called '*free electron*' which is now regarded as being responsible for the forces between atoms of an element. These electrons —one to each atom in sodium—in swinging from atom to atom, hold together the structure made up of nuclei surrounded by the inner shells. The free electrons, or *valency electrons*, are no longer attracted to any particular atom, and description of atoms in which specific distributions of electrons are given must be regarded as averages only. Thus a sodium atom will have its inner shells complete, and, on the average, there will be one electron in its outer shell, but it will not be the same electron for more than a vanishingly small period of time.

The force holding the atoms together is called the *metallic* force, the name implying that we are at a loss to give it any physical description. The atoms are held together because they share a common system of electrons and one cannot properly talk about a force between separate atoms. Strangely enough, the interaction is similar to that between

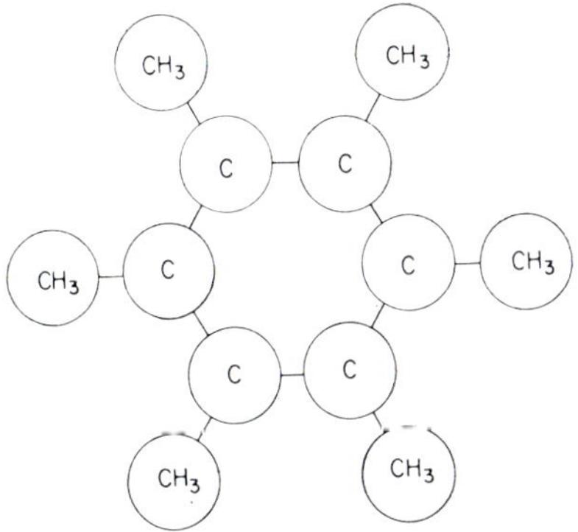

Fig. 9.1. The molecule of hexamethylbenzene.

atoms in what appear to be completely different materials—organic compounds—and it is worth while discussing the relationship at this stage.

In organic compounds we can distinguish a simple unit—the molecule. This is made of carbon atoms combined with others, particularly hydrogen, oxygen and nitrogen. Organic chemistry is based upon the idea of the molecule and the subject would have been hard hit if X-ray crystallography had shown that they did not exist. However, early structural studies confirmed all that the chemists had postulated—so much so that organic chemists tended to look upon the new methods as of little use since they told them only what was already known.

But in fact, they did tell more; they gave the spatial distributions of atoms in molecules and of molecules in crystals, and emphasized the three-dimensional nature of interatomic forces in place of what many people, influenced by diagrams in books, had come to believe was only two-dimensional.

Let us take hexamethylbenzene as an example; this was one of the first organic molecules to be worked out in detail and it showed clearly the benzene ring—a regular plane hexagon—with a CH_3 group attached to each corner (fig. 9.1). What holds this small group of atoms together?

The answer appears to lie in the electronic structure of carbon; it has two electrons in the K shell and four in the L shell. These four electrons cannot be regarded as free, giving carbon metallic properties, nor is the number just short of eight—the number in a completed L shell—to give ionic properties. Almost certainly, it is this delicate balance that gives carbon its particular properties, and which makes it the most important atom in the periodic table, since it forms the basis of living matter.

The accepted idea is that the force between two atoms represents an attempt to produce stability by sharing electrons to produce the desired number, eight, an octet. Thus each carbon atom will be happy to have four hydrogen atoms round it, each hydrogen atom contributing one electron (fig. 9.2). The hydrogen atoms are not regarded as becoming

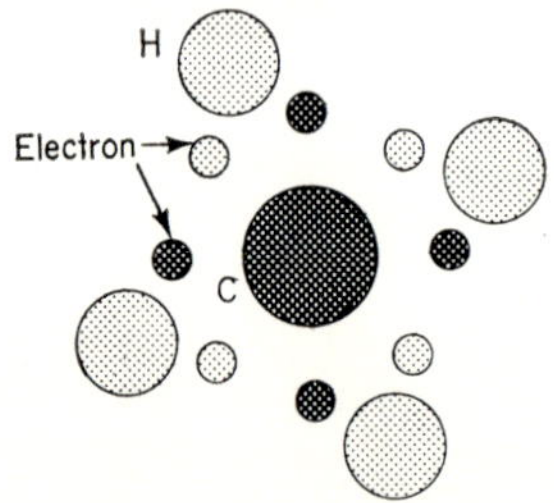

Fig. 9.2. The molecule of CH_4. The carbon atom and its four electons are heavily shaded and the hydrogen atoms, each with its one electron, are lightly shaded. The molecule is not in fact planar.

ions, however; each one is satisfied to have one of the carbon's four electrons associated with it, to form a stable group of two, corresponding to the inner K shell. Thus all the atoms share electrons; all contribute to the pool and all are well satisfied and have few other desires from the external world. The forces holding the group together are called *covalent* forces.

The simplest molecule that fits in with these ideas is methane, CH_4, but we have not used it as an example because it is not solid at ordinary temperatures; it is a gas. The molecule hexamethylbenzene, $C_6(CH_3)_6$, which we have introduced, is more complicated; each carbon atom in the methyl group CH_3 has a shell of seven electrons, and therefore requires another electron from the central six-fold ring to complete its octet. This requirement accounts for what is called the single bond of the CH_3 group.

The crystal structure of $C_6(CH_3)_6$ showed the molecules clearly, with carbon atoms about 1·5 Å apart. But the distances between carbon atoms in neighbouring molecules were much greater—about 3·3 Å. Since we have assumed that the atoms within the molecule are satisfied with their groupings, we have now to account for the force that makes the crystal hold together.

Since the intermolecular distances are much greater than the intramolecular distances, the forces between the molecules must be very weak. It has been the custom to call them *residual forces*—the forces that remain when all others have been eliminated. In particular they must be responsible for the crystallization of the rare gases, such as neon and argon, at very low temperatures. Since the force is that which van der Waals introduced to justify the constant a in his well-known equation:

$$\left(P+\frac{a}{v^2}\right)(v-b) = RT,$$

it is called the *van der Waals force*.

It is not satisfactory, however, merely to give it a name and leave it at that; naming a force does not explain it. There is no completely adequate explanation, but the most likely one is based upon induced electric moments.

One must not regard a molecule or even a single atom as a fixed invariable body; there can be fluctuations of electron distributions away from the equilibrium state. Fluctuations will probably lead to an electric moment and hence to an electric field surrounding the molecule. Any neutral molecule will experience an attractive force, just as an uncharged body is attracted to a charged one. Of course, the second molecule will also be fluctuating and any electric moment produced may cause a repulsion; but on the average the force produced by the fluctuations must be an attraction. This is supposed to be the origin of the van der Waals force.

The main forces in crystals are now accounted for, and there seems to be no reason for expecting any others. But, in fact, an anomaly is found to occur in some organic crystals; distances exist between atoms that cannot be accounted for by the forces so far described. In oxalic acid dihydrate, $(COOH)_2 \cdot 2H_2O$ for example, distances of 2·5 Å and 2·9 Å occur between oxygen atoms of different molecules; this is too large to represent an intramolecular force and too small to represent a van der Waals force. It is somewhere in between.

Examination of a number of crystals showed that this distance in organic crystals was always associated with the presence of a hydrogen atom, and the bond between the two atoms was therefore called the *hydrogen bond*. Once again, however, we must guard against the tendency to think that because we have named it we have therefore explained it. What is particular about the hydrogen atom that causes it to have this peculiar property?

Hydrogen is the lightest atom, consisting of a proton and an electron.

When it is ionized it loses its electron and is therefore the only positive ion that has no surrounding negative charges. It is believed that this is the property that is the origin of the hydrogen bond. The positive nucleus is exposed, and this can attract the negative electrons of atoms such as oxygen, nitrogen and carbon.

The hydrogen bond, in a way, is not a proper bond at all. It is rather weak—less than $\frac{1}{10}$th as strong as ionic and covalent bonds—and is therefore easily broken. On the other hand, it is much stronger than the van der Waals force, and compounds held together by hydrogen bonds have higher melting points than those which depend only on van der Waals forces.

Its weakness, however, is responsible for its great importance. Because it can be easily broken and re-made, perhaps between different atoms, it gives Nature the flexibility she needs to produce living matter; if changes could take place only by melting and re-solidification obviously life as we know it would be impossible. The hydrogen bond is responsible, for example, for the way the proteins grow, and the way in which they reproduce themselves. It is also the basis for more mundane things like some modern adhesives and even for the adhesion of dirt to our skins.

These are the main types of forces that we know. Although their existence was suspected before the discovery of X-ray diffraction, only the detailed investigations of crystal structures enabled them to be studied in detail. There is no reason, however, to suppose that the list is exhaustive; there must be other types of interatomic attractions as well. For example, the delicate process whereby a molecule of oxygen attaches itself to the haemoglobin molecule to enable us to breathe does not fit into any of these categories. There is still more to be found out about interatomic forces.

Also, we must point out that not all interatomic forces fall precisely into these various divisions; they may be partly electrovalent and partly covalent for example. The atoms in the hydrogen molecule, to take the simplest example, can be regarded as joined by an ionic bond, with one atom acquiring an electron from the other, or by a covalent bond with the two electrons shared by the two atoms. Thus we may say that any bond may have some ionic and some covalent character. Interatomic distances may even give us a rough estimate of the relative amounts of the two types.

Nor can we definitely assign every crystal to a particular class. Simple crystals such as NaCl are almost entirely ionic, but more complicated ones, such as the salts of organic acids, may have a mixture of all sorts of bonds. We shall give examples later, when we discuss the various types of bonds in detail, of crystals in which the different types coexist.

9.3 *The ionic bond*

So far, apart from the evidence leading to the hydrogen bond, only qualitative information has been needed to classify the different types of interatomic forces. Most science progresses in this way: one must understand a subject qualitatively first before trying to make it quantitative, and a premature introduction of mathematics may even delay ultimate understanding. But now we must see what quantitative information we can obtain from crystal structures, and we start with the ionic bond.

The simplest ionic structures are the alkali halides, typified by NaCl. What information can we obtain by comparing the sizes of the unit cell of the various salts? Clearly we can find the distances between the ionic centres; these are listed in table 9.1. It should be noted that the salts of caesium do not have the same structure as the other salts, but the interionic distances for caesium fit into a consistent pattern.

		Alkali atom				
		Li	Na	K	Rb	Cs
Halogen atom	F	2·01	2·31	2·66	2·82	3·00
	Cl	2·57	2·81	3·14	3·27	3·56
	Br	2·75	2·98	3·29	3·43	3·71
	I	3·00	3·23	3·53	3·66	3·95

Table 9.1. Interionic distances in the alkali halides.

This pattern shows that, as we change from one alkali ion to another, keeping a particular halide ion constant, there is more or less the same increase in inter-ionic distance. The same rule applies if we change the halide ion, keeping the alkali ion constant. The reader can check these facts from table 9.1; he will find that the results are consistent to within about 0·1 Å, which is satisfying enough at this stage. The meaning of this is clear: each ion can be allotted a definite radius, to a few hundredths of 1 Å.

Now that we are used to this idea, it does not seem as extraordinary as it did to the early workers, who did not expect anything so simple. Ions are complicated things and the possibility of being able to represent them by solid balls with specific radii seemed quite improbable. Nevertheless, this was how it turned out.

To find the actual radii was more difficult. The crystal structures give only the sums and not the separate values, and therefore quite different self-consistent sets of radii could be found. For some time it was thought that the metal ions were larger than the halide ions, but when more complicated structures were studied, with ions of the same sort in contact, this was found not to be so. A set of ionic radii, from a large number of crystals, could be drawn up, and the results were tabulated in an early paper by the German geologist Goldschmidt. They are shown in graphical form in fig. 9.3, which shows clearly the pattern of repetition as we transverse the groups of the periodic table. Goldschmidt's results have not been greatly altered over the years.

It is surprising how far the simple idea of ionic radius can take us. We know, for example, that in NaCl each ion lies in a regular octa-

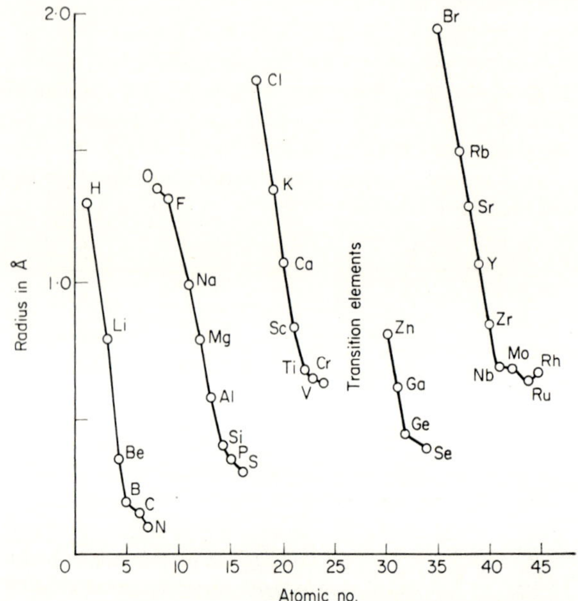

Fig. 9.3. Ionic radii of some of the elements.

hedron formed by six opposite ions. Since the section of an octahedron is a square, the NaCl structure requires the ratio of the radii of the two ions to be about $\sqrt{2}-1 = 0{\cdot}41$ (fig. 9.4). For NaCl, the measured ratio is fairly near $-0{\cdot}52$. Now, Cs is a larger ion than Na, and for CsCl the measured radius ratio is $0{\cdot}93$; thus this structure has

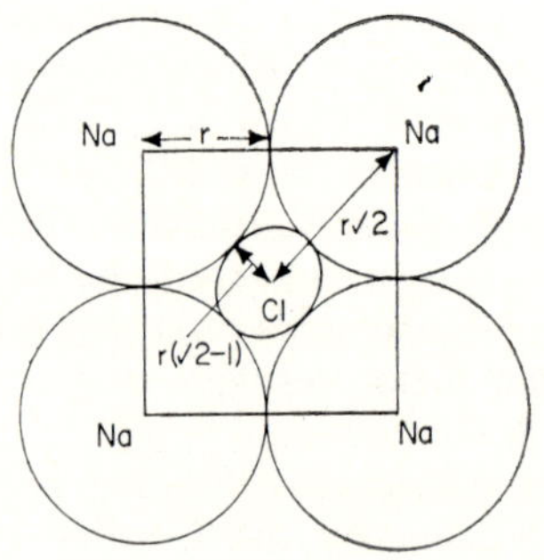

Fig. 9.4. Relative radii of the ions in NaCl.

to adopt another configuration ,with eight ions round each, requiring a radius ratio of about $\sqrt{3}-1$, or 0·73. Simple geometry, not much more complicated than this, plays a very large part in deciding the structures of chemical compounds.

It is clear that considerations of this sort will work only if the positive and negative ions are roughly the same size. If they are very different, Nature finds a way round the difficulty by clothing the smaller ion with

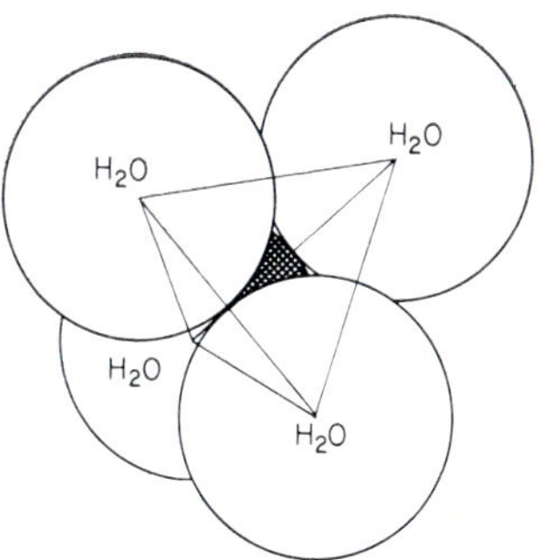

Fig. 9.5. Tetrahedron of four water molecules around the Be atom (heavily shaded).

molecules from which the crystals grow. For example, the sulphate ion, SO_4^{2-}, is much larger than the Be^{2+}; crystals of $BeSO_4$ therefore grow with four molecules of water surrounding the small Be^{2+} ion (fig. 9.5). The arrangement Be . $4H_2O$, called a *coordination group*, is similar in size and shape to the SO_4 group, and the two together form a beautifully symmetrical tetragonal structure.

There are not many ions as small as Be, and most metals cannot be accommodated in a tetrahedron of water molecules; six is a much more

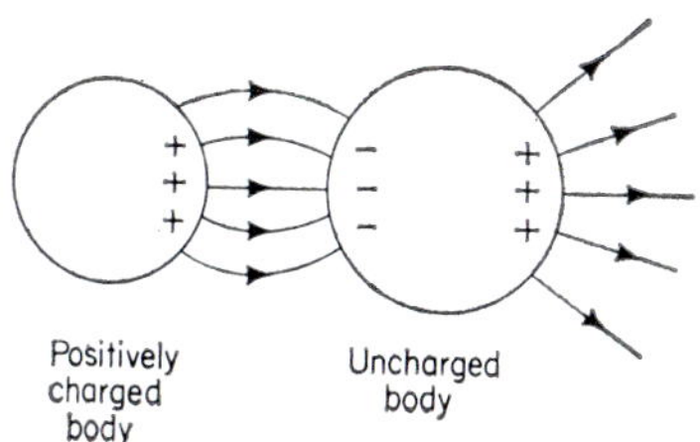

Fig. 9.6. Attraction between a charged and uncharged body.

usual number. Nickel forms sulphates with both $6H_2O$ and $7H_2O$; each has a octahedron of water molecules around the metal. It is again surprising how these simple considerations turn out to be so important.

But, of course, we cannot accept, without deeper consideration, the concept that the water molecules merely 'clothe' the metal ion; there must be some force between them that holds the composite ion together.

This is, in fact, the well-known polarization force that exists between a charged and an uncharged body (c.f. p. 129). From Coulomb's law, stating that the force is proportional to the product of the charges, we should expect that there should be no such force, but we know that it does exist. The presence of a charged body causes an attraction of the

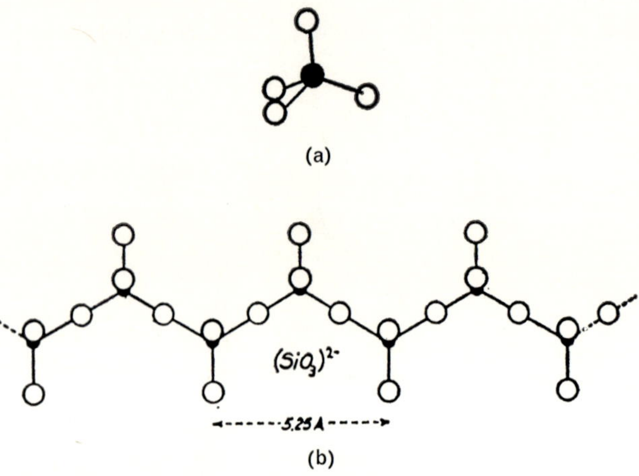

(a)

(b)

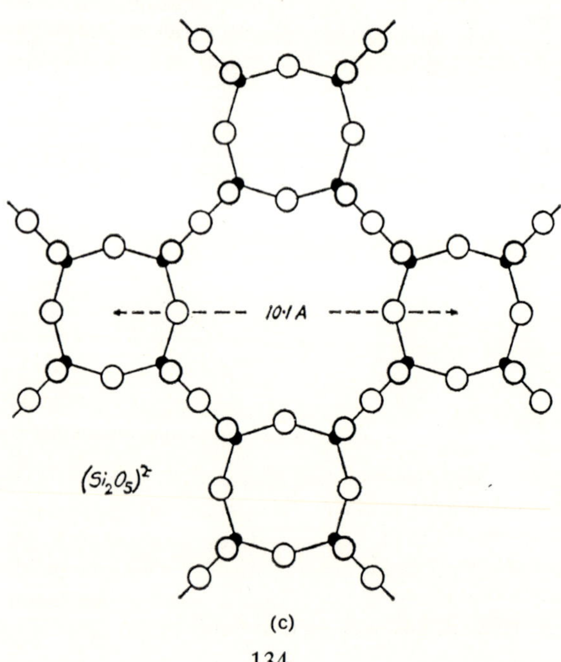

(c)

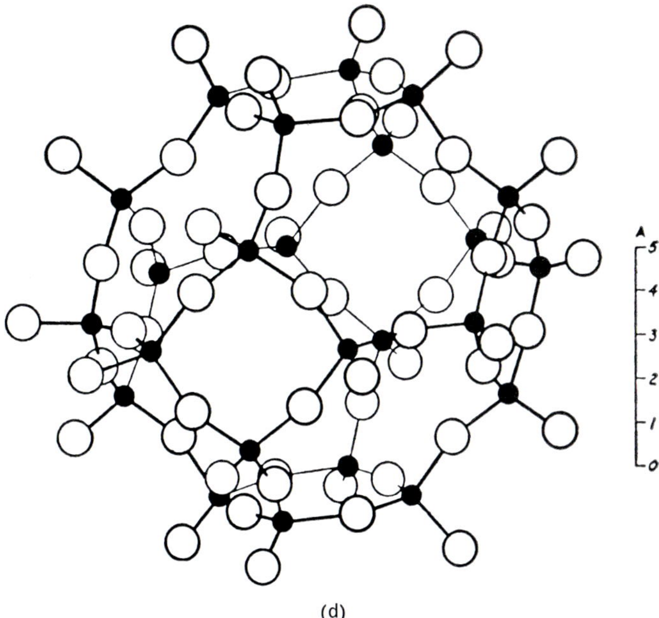

(d)

Fig. 9.7. Examples of different structural units possible in silicates: (*a*) SiO_4; (*b*) SiO_3 chain; (*c*) Si_2O_5 sheet; (*d*) $AlSiO_4$ three dimensional framework (the black spots represent Al or Si).

opposite charge on the nearer side of the uncharged body and repulsion of a similar charge to the further side (fig. 9.6); since the opposite charge is nearer, the resulting force is an attraction. The uncharged water molecules can be attracted to the positively charged metal ion in this way.

Charges, however, cannot redistribute themselves on an atom as easily as they can on a macroscopic body. In fact, they can do so only because the water molecule has a structure—two hydrogen atoms and one oxygen atom. Ammonia, NH_4, can operate in the same way. A single atom, such as one of the inert gases, could not behave in this way, and this is one of the reasons why these atoms do not readily enter into chemical combination with others.

It would take up too much space to discuss coordination compounds in detail—how copper sulphate manages with $5H_2O$, and what the extra water molecule does in $MgSO_4 . 7H_2O$, Epsom salts. Nature knows how to build stable structures, and we cannot yet compete with her by calculating theoretically which arrangements are in fact the most stable.

Silicates, which are important because they form a large proportion of the Earth's crust, are also held together mainly by ionic forces, the basic element being the SiO_4 ion. This is similar in shape to the SO_4 ion, but since Si is tetravalent, its charge is greater. Its behaviour, however, is quite different; for reasons not fully understood, silicon atoms can form ions of tremendous—one can almost say infinite—complexity. Whereas sulphur can form relatively simple ions like the persulphate ion S_2O_7, in which two tetrahedra share a common corner, silicon can form one-dimensional, two-dimensional and three-dimensional networks, running right through the crystals that they compose (fig. 9.7). It is for this type of reason that we have minerals with such a wide variety of properties—mica, which cleaves so beautifully (p. 28), and asbestos, which forms fibres, for example.

Ionic forces, because they are electrostatic in nature do not operate in particular directions from the atoms; they are not *directed forces*. Each ion collects around it as many ions as possible of opposite charge. The forces are very strong, and ionic compounds are characterized by their hardness and high melting points. NaCl, for example, melts at 800°C. These properties contrast with those in organic compounds, which are held together by weak van der Waals forces; they are therefore soft and have low melting points.

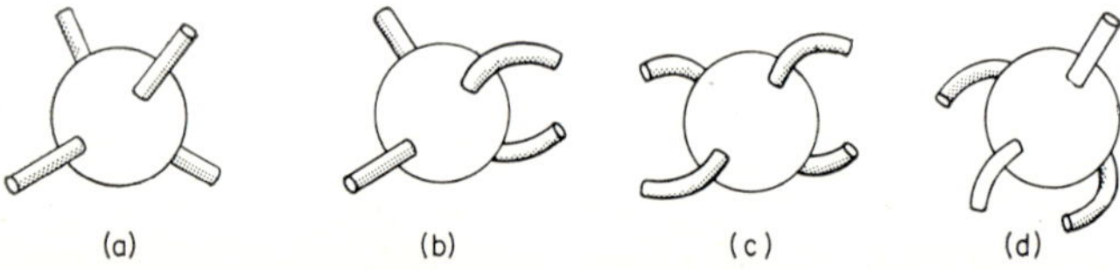

Fig. 9.8. Bonds in organic molecules. (*a*) The four tetrahedrally arranged single bonds to a carbon atom; (*b*) disposition of two single bonds and one double bond; (*c*) disposition of two double bonds; (*d*) disposition of a single bond and a triple bond.

9.4 *The covalent bond*

The covalent bond, being concerned with the sharing of electrons by a small number of atoms, is a directed bond. That is, each atom can have others joined to it only in specific relative directions. The carbon atom provides a good example of this principle: in general its four bonds are equally inclined to each other so that they have the directions associated with the lines joining the centre of a regular tetrahedron to the corners (fig. 9.8 *a*). All organic molecules with single bonds obey this principle.

If there are double bonds present—that is, two electrons are contributed by one of the atoms to the joint pool—then the three bonds take up a coplanar configuration, roughly at 120° to each other (fig. 9.8 *b*). A single and a triple bond, or two double bonds, are usually linear (figs. 9.8 *c* and *d*).

We can see that with this sort of limitation the geometry of organic molecules is fairly specifically fixed by the arrangement of bonds. Only if there are single bonds present is there possibility of rotation of one part of a molecule with respect to another. It is again surprising how much can be deduced from simple geometrical considerations of this sort.

The relationships between molecules, as we have seen, is decided by van der Waals forces and by hydrogen bonds; since the latter are much the stronger, organic crystals containing them are harder and have higher melting points than those that do not. But the hardnesses and melting points do not approach the values that ionic compounds possess.

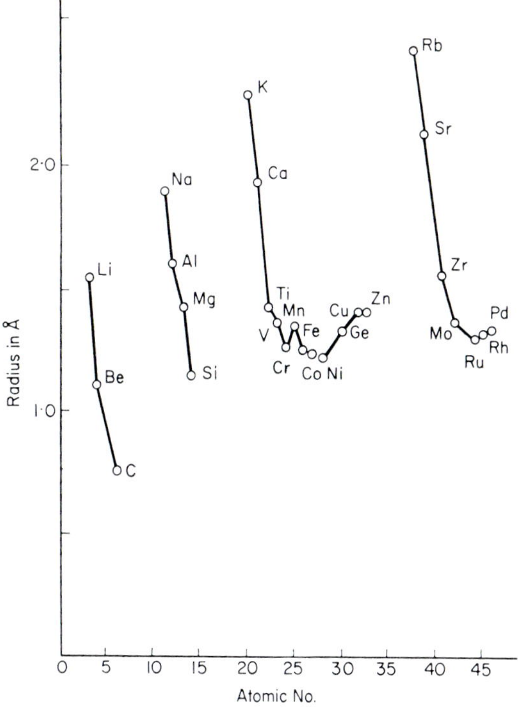

Fig. 9.9. Covalent radii of some of the elements.

A table can be drawn up of covalent radii of the elements and the results are shown graphically in fig. 9.9. It will be seen that covalent radii are much larger than ionic; it is surprising how much difference the omission of one or two loosely bound electrons makes. The carbon atom, being so important, has been investigated in great detail, and clear distinctions between the lengths of single, double and triple bonds have been found. Figure 9.10 gives a graph showing the relation between these lengths and from this it is claimed that one can, from the measurement of a specific bond length, deduce how much double-bond or triple-bond character it has.

Although the idea of the covalent bond as a sharing of electrons between two atoms has been extremely useful, modern ideas suggest that it is an over-simplification. It is thought that all the electrons that take part in bond formation are shared by the whole of a chemical molecule; in benzene, for example, whose molecule consists of a hexagon of carbon atoms with hydrogen atoms attached to each corner, we must consider the electron as performing complicated motions throughout the

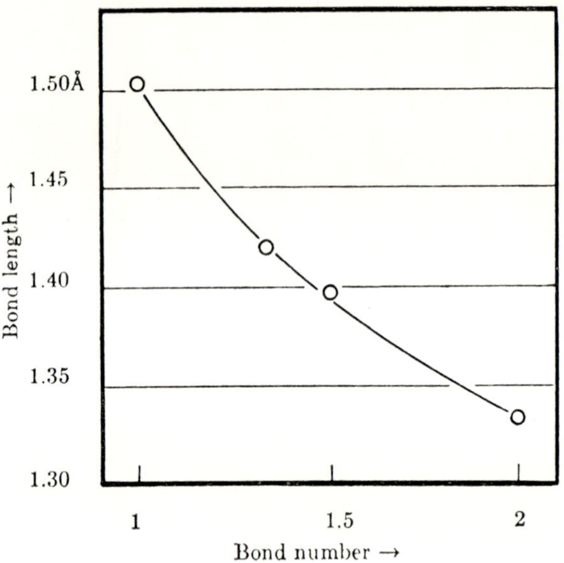

Fig. 9.10. Variation in length of carbon–carbon bonds, from the single bond to the double bond.

whole molecule. Their paths are called *molecular orbitals*. In effect, we can consider the molecule as a sort of complicated 'atom', containing its own electronic system with characteristic energy levels; because the 'atom' has a structure, the orbits are more complicated than for a single real atom. Only very simple molecules, however, can be treated theoretically in this way, and the covalent-bond idea is by no means out-dated.

9.5 *The metallic bond*

In effect, the metallic bond can be considered as the limit of the covalent bond, when the electrons are shared by *all* the atoms in a crystal, not just those in a molecule. In this way we can account for that most remarkable property of a metal—electrical conduction; in non-conductors, application of a p.d. causes displacement of electrons, but in metals it causes them to move and to continue in motion.

The radius of a metal atom is easy to measure, since one merely has to find the structure of the metal and measure the atomic separation; this must be the diameter. There is no complication involving the relationship between two different sorts of ions, as in inorganic salts (p. 133). It turns out that metallic radii are not greatly different from covalent radii, as we should expect since the forces are essentially similar in nature.

What is puzzling, however, is that the metallic force is not directed; most metal structures appear to be formed by packing atoms together as closely as possible (p. 89). The reason may be that the electrons responsible for metallic bonding are the so-called free electrons (p. 126) which are only loosely bound to the atoms; for covalent bonds more firmly attached electrons are involved and these cannot dissociate themselves completely from their orbitals within the atoms.

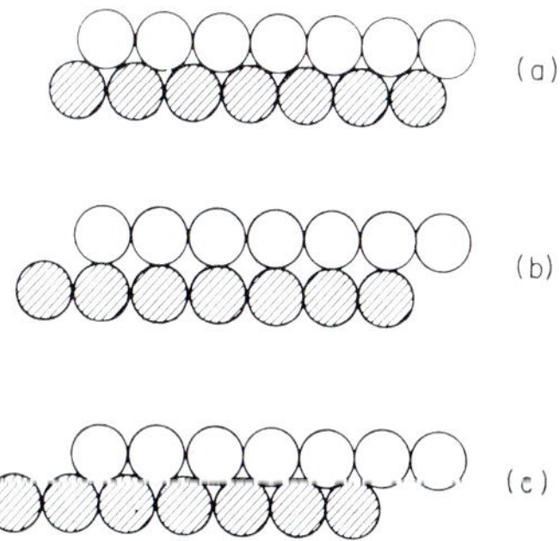

Fig. 9.11. One-dimensional representation of row of atoms (unshaded) moving from one position of stable equilibrium to another. (*a*) Initial positions; (*b*) intermediate positions; (*c*) final positions.

Because most metallic structures are formed of closely packed atoms, the planes of which they are composed can glide relatively easily over each other; the atoms can rise out of the depressions in which they are nestling, and all fall back together into the next depression (fig. 9.11). This simple idea accounts quite adequately for the ductility of metals. Most pure metals are very soft and ductile; single crystals of cadmium, for example, can be easily stretched into a ribbon by pulling by hand. For practical purposes, pure metals are never used; they are strengthened by the addition of foreign atoms. Thus brass—a copper–zinc alloy—is stronger than copper, and platinum for medals is alloyed with gold. Aluminium was not much use for kitchenware until a way was found of hardening it by small amounts of copper (see p. 182).

A delightful way of illustrating these points was introduced by W. L. Bragg in 1944. He formed a raft of equally sized bubbles on the surface of a soap solution, and they naturally formed the equivalent of a perfect crystal (fig. 9.12 *a*); he showed that the raft could be deformed easily, and

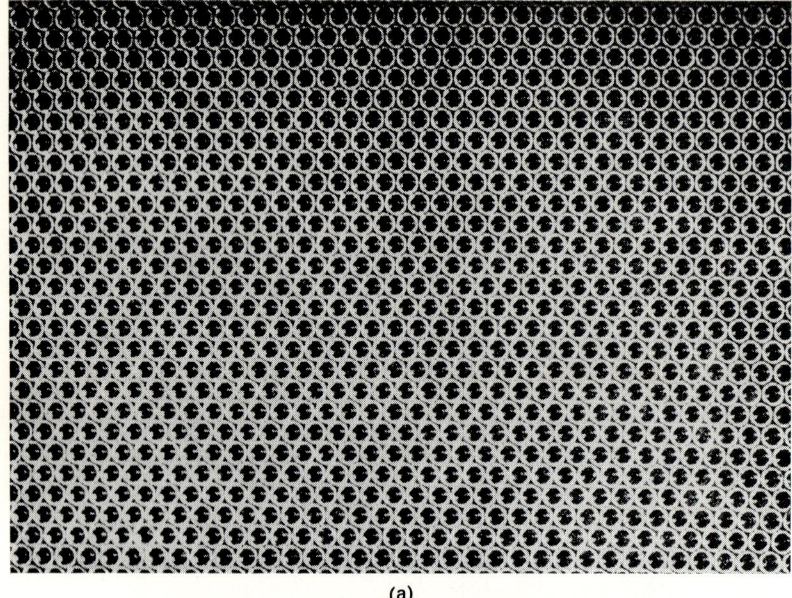

(a)

(b)

Fig. 9.12. (*a*) Raft of equally sized bubbles, representing perfect crystal; (*b*) raft with one bubble of different size, showing crystal imperfections introduced.

that the lines of bubbles would settle into new positions forming a perfect crystal again. But if a foreign bubble of different size were present (fig. 9.12 *b*), it would inhibit these movements, and greater forces would be necessary to cause changes.

These ideas lead us naturally to the subject of alloys—intimate mixtures of two or more types of atoms. What exactly happens when two different metals are melted together and allowed to solidify? The simplest result may be that the atoms distribute themselves at random on the structural sites, forming what is known as a *substitutional solid solution* (fig. 9.13 *a*). Alloys of copper and nickel, for example, are all face-centred cubic, the size of the unit cell decreasing slightly as the copper atoms are replaced by nickel.

For most alloy systems, however, there is a limit to the extent of this solid solution; when a certain proportion of foreign atoms has been

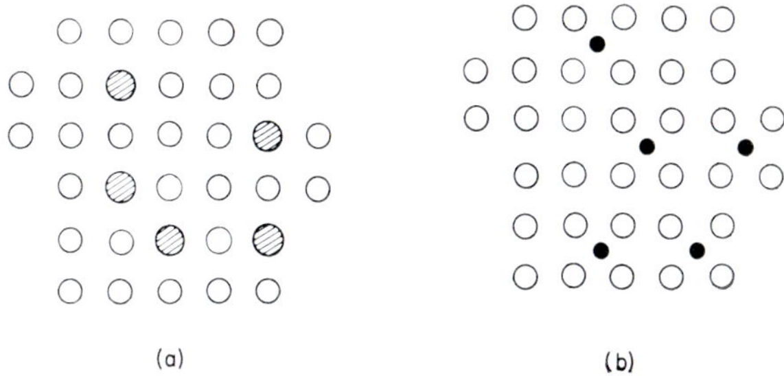

Fig. 9.13. (*a*) Substitutional solid solution; (*b*) interstitial solid solution.

accommodated, a new atomic arrangement begins to appear, and the alloy is then said to be *two-phase*. The second phase increases in extent as more foreign atoms are added, until a composition is reached at which only the new phase exists; this is called an *intermetallic compound*. Unlike a chemical compound, however, it is not usually fixed in composition. This succession of single-phase and two-phase regions may occur several times in a particular alloy system, and immense variety is provided by the many binary alloy systems that have been studied. With alloys of more than two metals the variety is considerably greater.

Intermetallic compounds, being composed of atoms of different sizes, are sometimes rather complicated; we have already discussed one such structure—γ-brass, Cu_5Zn_8 (p. 90). What is surprising, however, is that some elements have complicated structures also. Why should atoms, presumably all of the same size, choose to pack together in a rather irregular way? Manganese is an example: it has several possible forms, associated with different temperature ranges; α-manganese, stable at room temperature, has a cubic unit cell containing 58 atoms.

It is possible that manganese is composed of two sorts of atoms, with different distributions of electrons. Most textbooks, when dealing with the subject of electron distributions in atoms, make statements that are far too definite. For example the electron distributions of the elements with atomic numbers from 21 to 30 are usually given as follows

Element	At. No.	Electron shells			
		K	L	M	N
Sc	21	2	8	1	2
Ti	22	2	8	2	2
V	23	2	8	3	2
Cr	24	2	8	5	1
Mn	25	2	8	5	2
Fe	26	2	8	6	2
Co	27	2	8	7	2
Ni	28	2	8	8	2
Cu	29	2	8	10	1
Zn	30	2	8	10	2

These must be taken 'with a pinch of salt'; clearly Cr and Cu are exceptions in the sequence. Mn could be exceptional also, and it may be possible that it has some atoms with six electrons in the M shell. The complicated α structure may therefore be regarded as an intermetallic compound with two types of atom.

Not all solid solutions are substitutional; if the second atom is very small it may fit into the spaces between the atoms of the first type, forming an *interstitial solid solution* (fig. 9.13 *b*). By far the most important example of this is carbon in iron and this is responsible for the properties of steel, which is a material that has played a large part in the development of modern civilization.

9.6 *Solid-state physics*

The immense variety of structures of intermetallic compounds inspired certain people to try to find some general rules of behaviour of the elements when they combine with each other. Out of the first tentative empirical ideas new principles arose, and new theories were developed. Now the subject of solid-state physics has come into being, and is already one of the most important branches of physics, with a direct impact upon our lives—if only because it has produced the transistor.

One of the first people to interest himself in alloy problems was Hume-Rothery of Oxford, in the 1920's; he put forward the idea that one of the important quantities deciding the structure of an alloy was the ratio of free electrons to atoms. The number of free electrons contributed by each atom was normally the valency; Cu had to be regarded as monovalent, Zn divalent, Al tervalent and so on. The transition elements—such as Fe, Co, Ni—had to be regarded as contributing no electrons; the explanation was that the unfilled inner shells absorbed as many electrons as the atoms contributed.

We may take the γ-structure (p. 90) as an example. It, or a close approximation to it, is formed in several alloy systems and the following table shows that they have the remarkable ratio of electrons to atoms of 21/13.

Alloy	No. of Atoms	No. of electrons
Cu_5Zn_8	13	$5+2\times 8 = 21$
Cu_9Al_4	13	$9+3\times 4 = 21$
$Cu_{31}Sn_8$	39	$31+4\times 8 = 63$
Fe_5Zn_{21}	26	$0+2\times 21 = 42$

Some other types of structure showed the same tendency. What did it mean?

The complete story is too complicated to tell here, but it is something like this. The electrons in a crystal obey the same rules as those in an atom; they have energy levels which only certain numbers of electrons can occupy. Because large numbers of atoms are involved, the energy levels are very close, but nevertheless, because there are so many electrons, some of them have very high energies. These are the free electrons (p. 126) which must be pictured, even at absolute zero, as dashing about in a crystal with considerable velocities.

Now, a moving electron has a wavelength λ, defined by de Broglie's principle—$\lambda = h/mv$, where h is Planck's constant, and mv is the momentum. These waves can interact with the lattice in accordance with Bragg's law, in the same way as X-rays. It is this interaction which is held to be responsible for the importance of Hume-Rothery's electron–atom ratio.

Some of the electron waves could have the right direction and the right wavelengths to be reflected by the lattice planes (p. 47). Then they would be reflected to other directions by these planes, but it can be seen from fig. 9.14 that they would still be reflected from the same planes. In fact, theory shows that such electron paths cannot exist. If we increase the electron content by adding a metal of higher valency to one of lower valency, when we reach compositions near to those that would produce such electrons, interesting happenings occur.

First, the electron energies become less than those which would correspond to the quantity $\frac{1}{2}mv^2$. Secondly, their directions may change so that fewer of them approach to the Bragg condition. Now, we know that the condition for stability of a system is that it should have minimum energy, and it is the reduced energy of the electrons that is supposed to be responsible for stabilizing the structure. A complicated structure may have several different planes that reflect, and so produce more possibilities of stabilization.

As more electrons are added, they do not want to break through the forbidden band of energies because this would produce a discontinuous increase; they therefore tend to adopt paths that allow the smaller

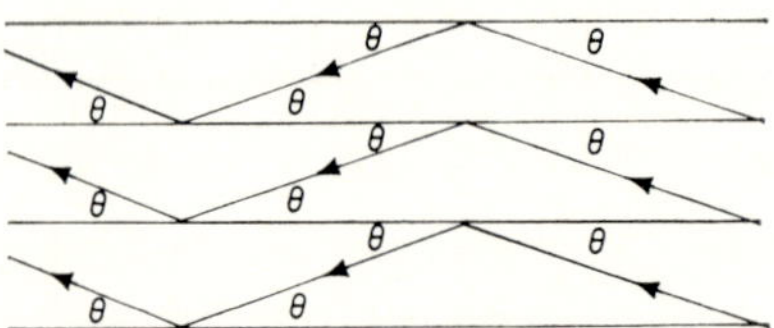

Fig. 9.14. Electron waves reflected from lattice planes at angle θ are incident at angle θ on the same planes.

increase that results without breaking through. But if the energy increases so much that a lower energy would result by penetrating the barrier at its lowest point, then this penetration will occur. This is the behaviour that characterizes a conductor; the energy gap varies with

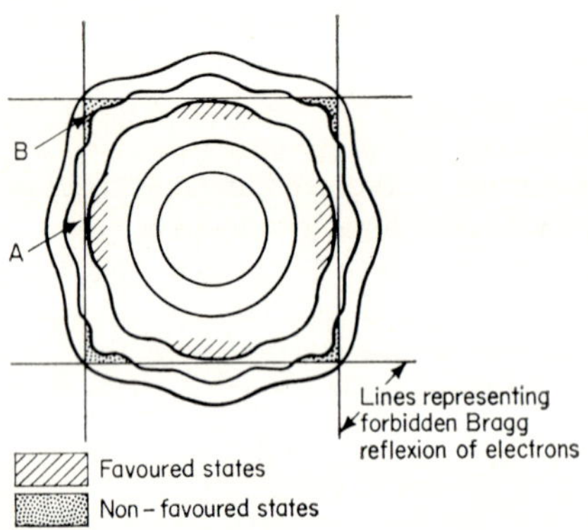

Fig. 9.15. Two-dimensional representation of energy E of electrons as function of momentum p. Near the centre E is equal to $p^2/2m$, and so the contours are circular. Further out, E is dependent also upon direction of momentum, and some directions lead to lower energies and are therefore favoured. A region such as A, outside the electron barrier, may have a lower energy than a region such as B, which is within the barrier.

direction in such a way that the highest values on the lower side (fig. 9.15) are higher than the lowest values on the upper side. In other words, there is no absolute gap

If, however, there *is* an absolute gap, the electrons cannot break through, and the crystal is an insulator. If there is a very small absolute gap heat energy may cause electrons to bridge the gap and so produce some conduction; this happens in semi-conductors, such as Si and Ge, which have the property that their resistance decreases with temperature. For ordinary metals, resistance increases with temperature because the motion of the atoms increases the imperfection of the crystalline array.

All these happenings occur naturally. But the transistor is not natural. It has been found possible to insert small numbers of impurity atoms in semiconductors so that they provide stepping-stones across the energy gap. The impurity atoms must be of valency 3 or 5—that is, one more or one less than that of the host semiconductor. Then minute control can affect the behaviour of electrons and alter the way in which the whole crystal responds to external electrical stimuli.

Those who know the subject will realize that we have been describing what is known as *Brillouin-zone* theory. We have thought it best, however, not to introduce more technical terms than necessary, particularly since the full theory can be adequately dealt with only in three dimensions, with rather complicated diagrams. We have attempted only to give an indication of the way in which research in one field can open up approaches in others, and ultimately produce ideas quite unrelated to the original subject.

9.7 *Summary*

We have tried to show in this chapter the consequences of our increasing knowledge of the ways in which atoms organize themselves in crystalline matter. We can now classify chemical compounds, alloys and minerals, and can make reliable deductions about the types of interatomic forces that join the atoms to each other. The general outlines of this scheme of forces look sharp and clear; we see ionic, covalent, metallic, van der Waals and hydrogen bonds. But when we look more closely we see how these bonds merge into each other, and we are beginning to see more clearly the relationships between them. Moreover there are still some forces that we do not fully understand.

Perhaps the most surprising result of the investigation of crystal structures is the part played in them by simple geometrical considerations. The structures of some compounds—for example, the alkali halides—seem to be decided by the relative sizes of the atoms, pictured as hard solid spheres. The extent of solid solution of one metal in another is also largely decided in the same way. Phases may be precipitated in alloys merely because their structures have planes of atoms that fit neatly on to planes in the parent structure. It is even probable that shapes of chemical compounds are important in deciding their properties: the now-notorious DDT is a member of a group of isomers—compounds with identical chemical formulae, but different configurations of atoms; since it is the only one with its particular pesticidal properties, its molecular shape must be the deciding factor.

It may turn out ultimately, however, that the main impact of X-ray crystallography on society has occurred through the medium of solid-state physics. This subject owes more to the early X-ray diffractionists than present-day text-books indicate. The key concept is the diffraction of the valency electrons by the crystal structure in which they exist, and this concept was established by early work on solid solutions and on relatively complicated structures such as that of γ-brass. Semiconductors are elements in which the resulting electrical properties are finely balanced, and so can be tipped one way or another by the presence of small numbers of atoms of different valency.

It can be seen, then, that the contribution made by X-ray diffraction to the knowledge of the nature of solid matter is immense. We may liken this contribution to that made by modern surveying instruments and aerial surveys in map-making. Ancient maps were made from a few arduous journeys, supplemented by a great deal of guesswork and imagination; they necessarily contained many mistakes and errors, and were quantitatively almost useless both in direction and distances. Modern maps can be relied upon completely, often to a higher degree of accuracy than we need.

In the same way, X-ray crystallography has brought into sharp focus much that was vague and shadowy in our knowledge of matter. Mistakes have been corrected, detail has been filled in, and qualitative ideas have been made quantitative; clear courses can now be charted. The worker in any branch of science who wishes to explore any particular section of it now has a clear and accurate map to guide him.

CHAPTER 10

comparison of X-ray and optical diffraction

10.1 *General principles*

IN Chapter 4 we have explained the relationship between the diffraction of X-rays by a crystal and the diffraction of light by a grating. In the present chapter we shall explore this comparison further. There are two reasons why we think that this comparison is important. First, it explains the basis of the subject by relating it to other branches of physics and so can be said to provide a deeper understanding than that provided by the mere manipulation of formulae. In X-ray diffraction, as in many other branches of science, it is possible to obtain significant results without a thorough understanding of basic principles, and it is one of the purposes of books such as this to discourage such superficial knowledge. Secondly, acquaintance with basic principles can often lead to new ideas; most of the ideas described in Chapters 6 and 8 were introduced by people with a thorough knowledge of what they were doing.

This philosophy can be looked at in another way. There are many who believe that, because physics is a quantitative subject, it must be based upon mathematics; therefore, the earlier that mathematics is brought into play, and the more extensively it is used, the better. This is not necessarily true. Mathematics is of no use if physical principles are not understood. The ability to follow the mathematical basis of a subject, and to manipulate the formulae so derived, is no substitute for sound physical understanding. The physics should always be mastered first, even if only qualitatively; mathematics can then be brought in to make the results quantitative. Let us look at X-ray diffraction from this point of view.

The basic idea is that the theory of X-ray diffraction by a crystal is simply an extension of diffraction-grating theory. A diffraction grating is a periodic repetition of a unit—a scratch on a glass or metal plate. The single unit has its own diffraction pattern, which is a continuous function; this is called the *diffraction function* (fig. 10.1 *a*). The effect of the repetition of the unit is to allow diffracted light to occur only at specific angles, θ_n, given by the equation:

$$n\lambda = d \sin \theta_n.$$

This set of angles is called the *interference function* (fig. 10.1 *b*). The complete diffraction pattern is the product of the two functions (fig.

10.1 *c*): the spacing decides where the orders will occur; the structure of the unit decides what their relative intensities will be.

10.2 *Optical apparatus for studying diffraction*

When a light wave from a point source falls upon an object, the form of the wave is altered, and the characteristics of the object are impressed

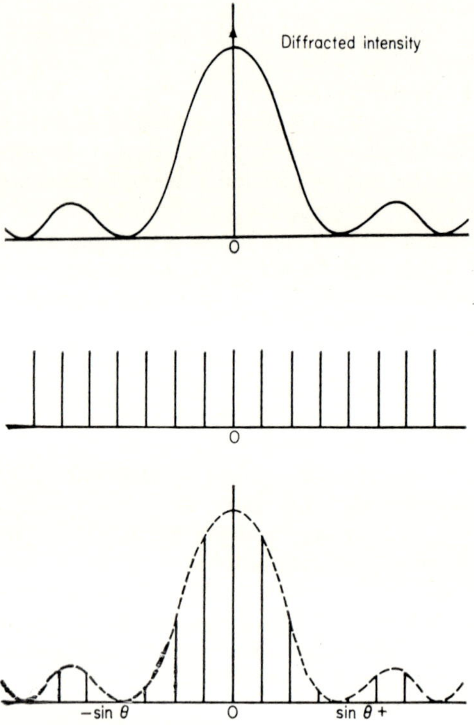

Fig. 10.1. Influence of unit of diffraction grating upon complete diffraction pattern. (*a*) Diffracted intensity as function of sin θ; (*b*) values of sin θ satisfying the diffraction-grating equation; (*c*) the product of (*a*) and (*b*) showing the complete diffraction pattern of the grating.

upon it (fig. 10.2). The resulting pattern of light is called the *Fresnel diffraction pattern* of the object. As the point source moves farther away from the object, the pattern changes; the extreme case, when the point source is at infinity, produces what is called the *Fraunhofer diffraction pattern.*

Diffraction patterns are easy to produce. All that is required is a strong source of light falling upon a pin-hole, and a screen, some metres distant, for observing shadows of objects produced by light passing through the pin-hole. The edges of these shadows will be seen to have fine fringes round them; these are called diffraction fringes.

To produce the special case of Fraunhofer diffraction is not so easy. Infinity is a large distance, and therefore to obtain the same effect we use a lens; if the pin-hole is at the focus, a plane wave will result. This plane wave can be focused again with another lens (fig. 10.3 *a*). If we put an object between the lenses, we see a Fraunhofer diffraction pattern surrounding the strong spot at the focus of the second lens.

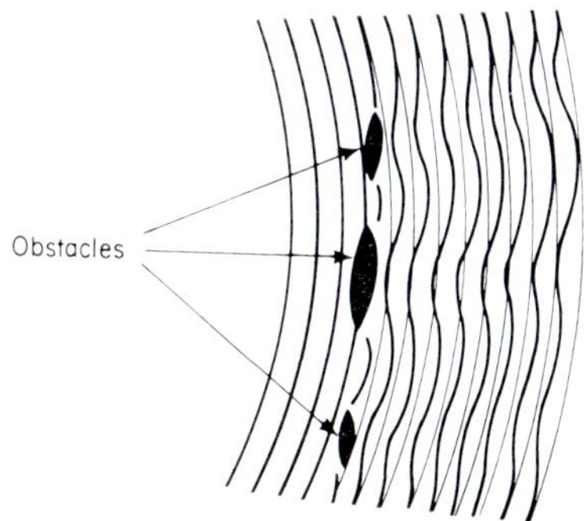

Fig. 10.2. How an obstacle, denoted by black areas, distorts a wave front.

The apparatus shown in fig. 10.3 *b* has been specially made for producing Fraunhofer diffraction patterns; the purpose of the mirror at the bottom is to present the pattern in a suitable direction for viewing and also to allow the observer to be near enough to move or change the diffracting object at will. The lenses have focal lengths of about $1\frac{1}{2}$ m, and so the whole apparatus is about 3 m high. It is called an *optical diffractometer*.

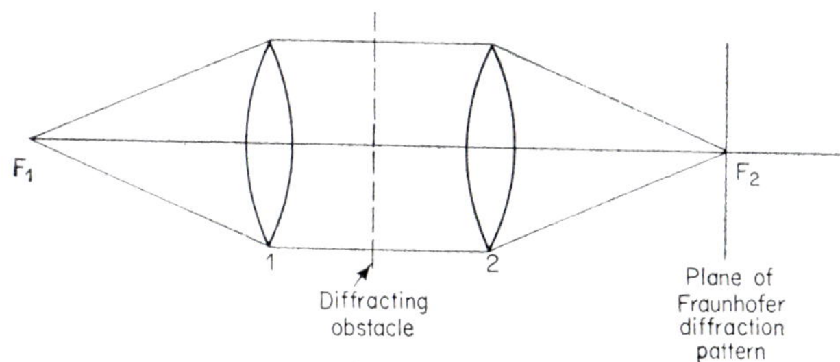

Fig. 10.3. (*a*) Production of Fraunhofer diffraction patterns. F_1 is the focus of lens 1; F_2 is the focus of lens 2.

As physics apparatus goes, it is very simple. Nevertheless it has some exacting requirements if the best results are to be obtained. First the lenses must be very good; doublets for astronomical telescopes, corrected for spherical aberration, will do if they are accurately made of good quality glass. They must be precisely coaxial—a requirement more difficult to fulfil than might appear. Considerable knowledge of the theory and use of lenses is required. (Geometrical optics is now generally held to be a dull and rather trivial subject, but it is nevertheless still a highly important one in many branches of physics.) Finally, the whole apparatus has to be strongly made so that it cannot easily be

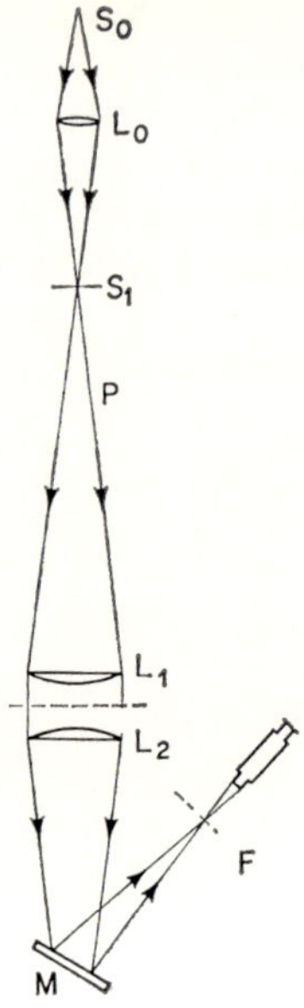

Fig. 10.3. (*b*) the optical diffractometer. The lens L_0 produces an image of the source S_0 on the pin-hole S_1, at the focus of the lens L_1. The lens L_2 focuses the parallel light, after reflection at the mirror M, at the point F. The diffraction pattern of an object between L_1 and L_2 is produced in the plane through F.

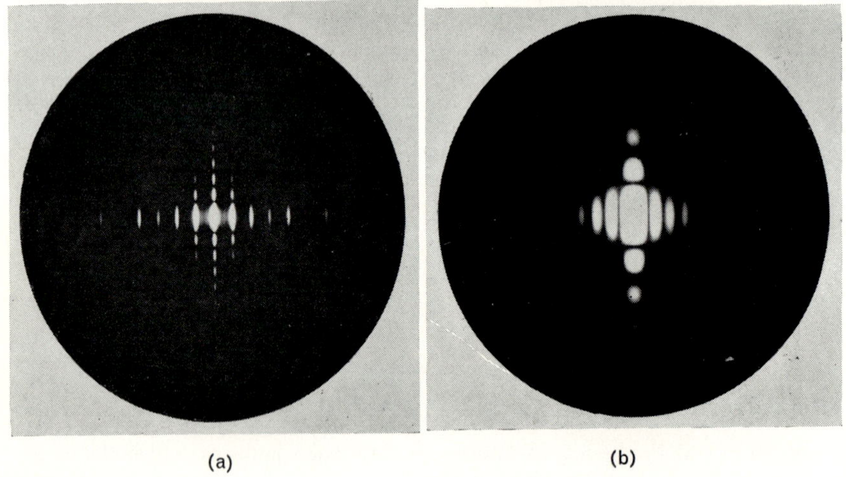

Fig. 10.4. (*a*) Diffraction pattern of rectangular hole; (*b*) diffraction pattern of grating with rectangular hole as unit.

deformed. The apparatus used to supply the photographs for this chapter was based upon a steel girder.

With this apparatus good Fraunhofer diffraction patterns can be obtained from objects several centimetres across. For example, fig. 10.4 *a* shows the diffraction pattern of a single rectangular hole $2\frac{1}{2}$ mm × 5 mm, and fig. 10.4 *b* shows that of a grating with this rectangle as its unit. Figure 10.4 *a* represents the diffraction function (p. 147) and the complete diffraction pattern can be seen to be based upon this.

The diffraction patterns illustrating this chapter were obtained on fine-grained photographic film placed in the focal plane of the second lens. Photographic film is very inferior to the human eye in accommodating itself to a range of intensities; with normal exposure and development, either the weaker parts of the patterns would not show or the stronger parts would be greatly over-exposed. We have therefore given very long exposures, and have developed for very short times. This method shows all the detail in which we are interested but the relative intensities are, of course, quite wrong.

A simple version of the apparatus is easily built and the reader is strongly recommended to look at some diffraction patterns for himself. Diffracting apertures can be easily cut with a razor blade in black paper or thin card.

10.3 *Illustrations of optical diffraction*

We may first ask why we are interested in Fraunhofer diffraction when the distance of the crystal from the X-ray source is certainly not infinite; it is normally about 50–100 mm. The answer is that the distance should be measured in wavelengths, and since the wavelength of X-rays is less than a thousandth of that of light, a distance of 50 mm with X-rays corresponds to over 100 m with light—near enough to infinity for practical purposes.

With the apparatus described in the last section, we can illustrate many of the phenomena of X-ray diffraction but usually only in two dimensions. We can make diffracting objects by punching holes in opaque cards, each hole representing an atom in the projection of a crystal structure on to a plane. This limitation is not severe; many structure investigations start with two-dimensional projections (p. 99) and, as far as principles are concerned, two dimensions serve as well as three.

As an example, let us see how the reciprocal lattice (p. 51) arises. Figure 10.5 *a* shows the diffraction pattern of a single hole, representing one point of a lattice. If we then punch a second hole, the diffraction pattern is modified by the presence of fringes—Young's fringes—perpendicular to the separation of the holes (fig. 10.5 *b*). If we now add another similar parallel pair of holes, giving the unit cell of the lattice, the pattern in fig. 10.5 *b* is crossed by a further set of fringes, which divide the basic diffraction pattern into spots (fig. 10.5 *c*). The two sets of fringes have spacings inversely proportional to the corresponding separations, and are perpendicular to the directions of the separations; these are properties of the reciprocal lattice that we noted on p. 51.

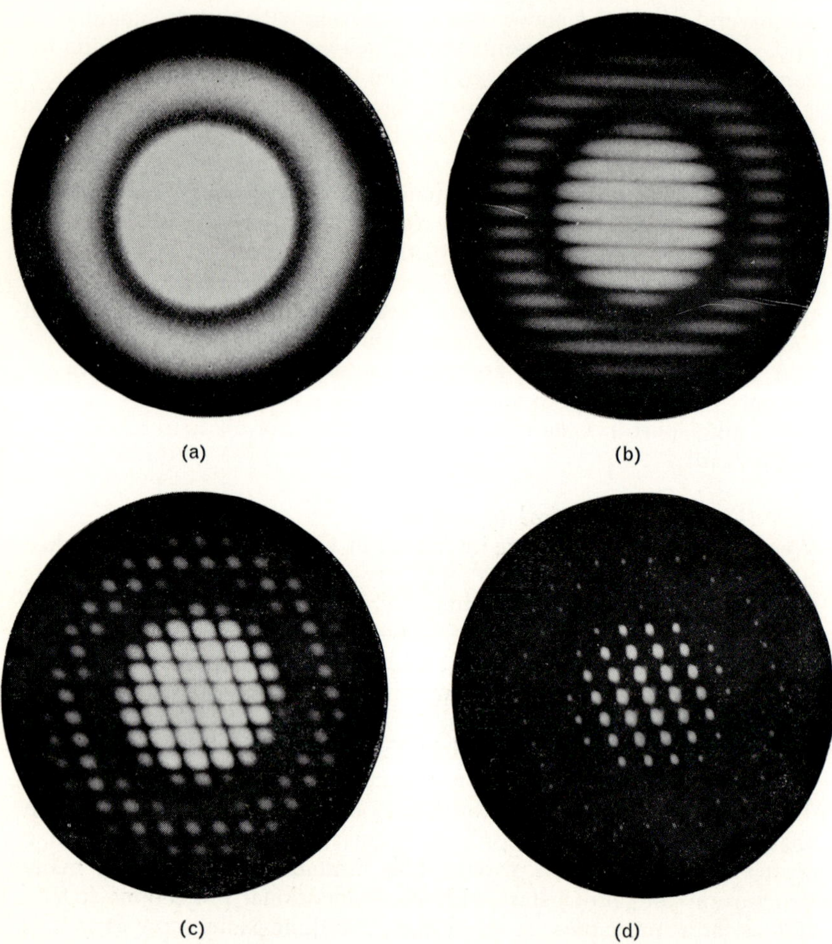

Fig. 10.5. (*a*) Diffraction pattern of a circular hole; (*b*) diffraction pattern of two similar holes (*a*), showing Young's fringes; (*c*) diffraction pattern of four holes (*a*) at corners of a parallelogram, showing crossed fringes; (*d*) diffraction pattern of two dimensional grating based upon (*c*).

As holes are added to produce a more extensive lattice, the spots maintain their positions but become sharper (fig. 10.5 *d*). If the lattice became infinite, the spots would become perfectly sharp. This illus-

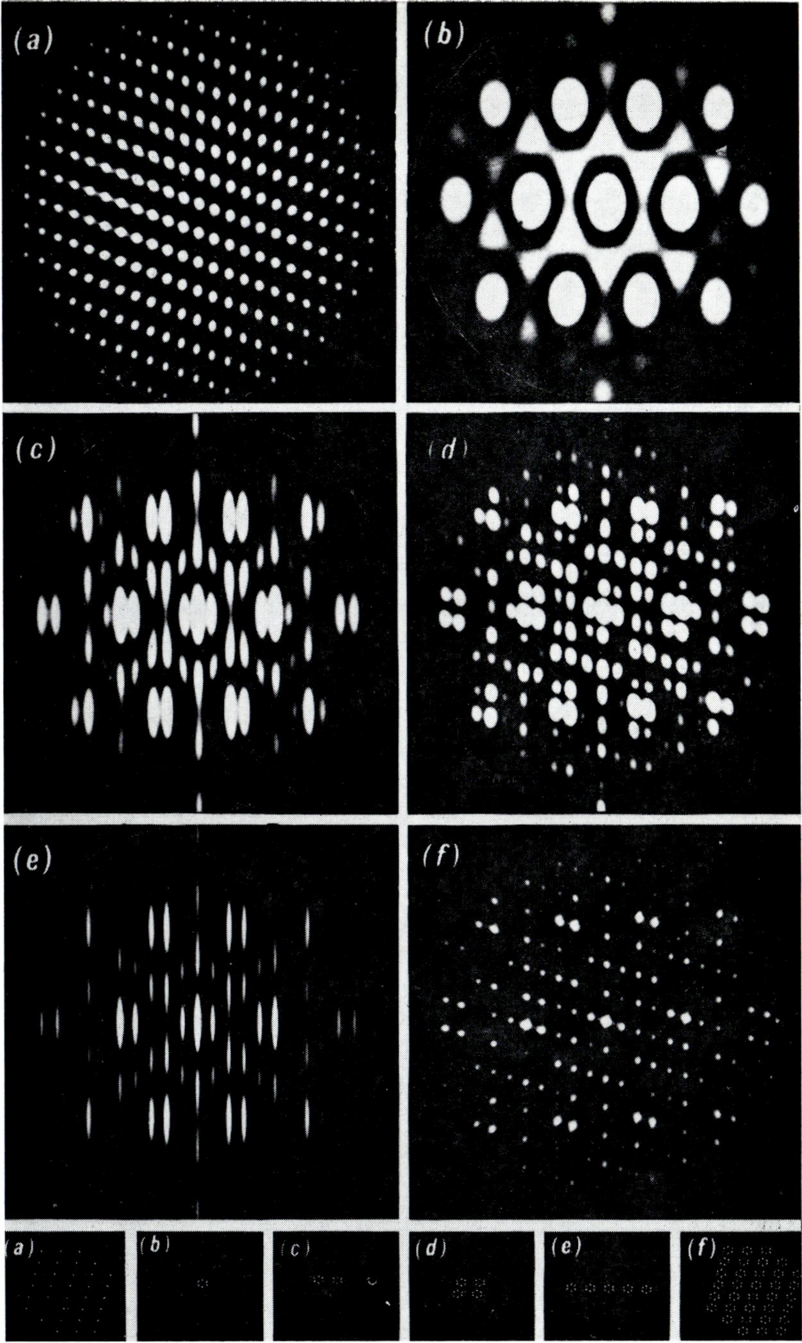

Fig. 10.6. Diffraction patterns of sets of small holes representing (*a*) a lattice; (*b*) the projection of a benzene ring; (*c*) two parallel sets of units (*b*); (*d*) four parallel sets of units (*b*), forming a unit cell; (*e*) a row of units (*b*); (*f*) a lattice, similar to (*a*), of units (*b*).

trates an important effect—that the sizes of spots given by small crystals increase as the crystals become smaller; the effect is appreciable, however, only with dimensions of about 100 Å or less and is therefore of value in measuring crystallite sizes only in a narrow range. We shall discuss this subject in more detail in Chapter 12.

We can make the equivalent of a crystal by repeating the experiment, this time replacing the single hole by a set of holes representing the projection of an inclined chemical molecule. The results are shown in figs. 10.6 *a–d*. Now we see that, instead of the steady fall-off in intensity from the centre outwards, the intensities are irregularly distributed between the reciprocal-lattice spots. This illustrates the effect of the structure factor (p. 58) on the intensity; as we have seen, the structure factors of neighbouring reflections in the reciprocal lattice can be quite different from each other.

We can see, then, that all the information about the diffraction pattern of a repetitive distribution of holes is contained in the diffraction pattern of the single unit; the effect of the regular repetition is to allow us to observe this basic diffraction pattern only at the reciprocal-lattice points—to 'sample' it, as it were. The importance of the basic diffraction pattern is emphasized by giving it a name—the *optical transform*. We can see that the optical transform is the diffraction function and the reciprocal lattice is the interference function as we have defined these terms in § 10.1.

10.4 *Uses of optical transforms*

The principles just described are not only of value in understanding the subject of X-ray diffraction; they suggest ways of tackling some of the problems discussed in earlier chapters. The subject is too large to be discussed fully here, and only one or two approaches will be described.

First, suppose that we wish to use the trial-and-error method of working out a crystal structure (chapter 6). We guess the positions of the atoms and see whether they give the right intensities. The optical-transform method allows us to do this with the minimum of trouble. We make an optical transform of the unit-cell contents and superimpose the reciprocal lattice upon it; we can then see whether the intense and weak regions of the transform agree with strong and weak X-ray reflections respectively.

For crystals that contain equal atoms—organic compounds, for example—the method works very well (fig. 10.7). Even if the guess is not quite right, the transform will tell us whether small adjustments are possible, for a peak in the transform must always lie near to a reciprocal-lattice point representing a strong reflection. No method of computation has this advantage.

Another use is to find the orientation of a plane molecule by producing its optical transform and seeing whether it can be changed in such a way that it fits correctly on to the weighted reciprocal lattice (p. 110).

An example is shown in fig. 10.8. It is surprising how accurately such a procedure can be carried out.

Finally, we can use the weighted reciprocal lattice to give the orientation of parts of a molecule. For example, the molecule of bishydroxydurylmethane has two hexagonal carbon rings inclined steeply to each

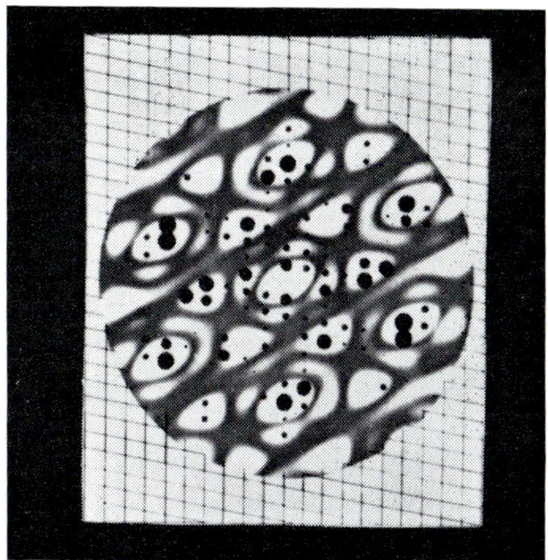

Fig. 10.7. Representation of the weighted reciprocal lattice of pyrene, $C_{16}H_{10}$, superimposed upon the optical transform of the unit-cell contents. The correspondence between the X-ray intensities (black spots) and the variations in the intensity of the transform is quite clear.

other, like the partly open wings of a butterfly. In the weighted reciprocal lattice (fig. 10.9) there are six sets of strong peaks, marked A–F, amongst the high-order reflections. These form a distorted hexagon, from the dimensions of which the projections on (010) of the two hexagonal rings in the molecule can be deduced; these projections are not

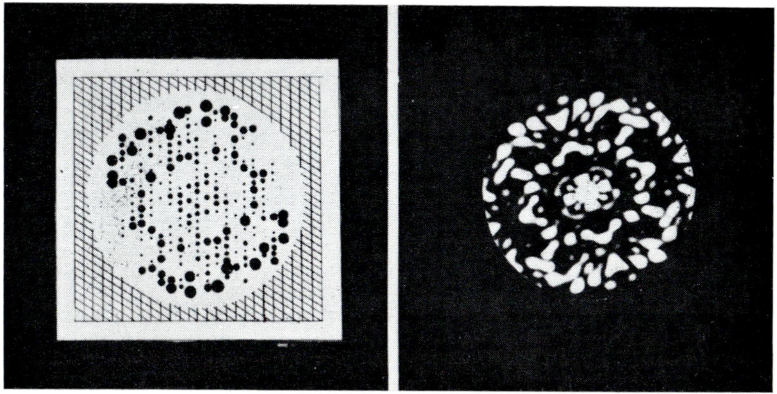

Fig. 10.8. Correspondence between (*a*) the weighted reciprocal lattice of phthalocyanine and (*b*) the optical transform of the appropriately tilted molecule.

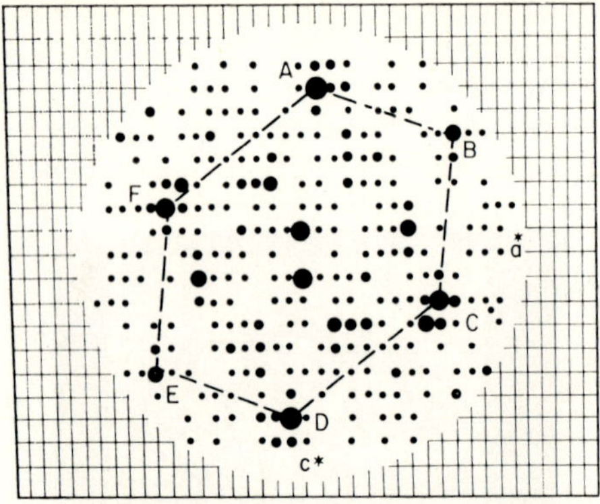

Fig. 10.9. The weighted reciprocal lattice of bishydroxydurylmethane, showing strong regions resulting from the hexagonal carbon rings.

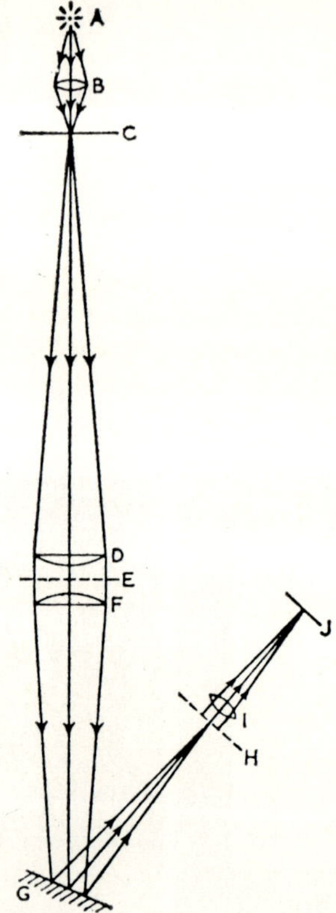

Fig. 10.10. Modification of the optical diffractometer (p. 150) to permit the production of the image of the mask. The components A–H are standard; the lens I produces an image, in the plane J, of the object E.

regular because the hexagonal rings in the molecule are foreshortened when viewed along the [010] direction. From this information, and the fact that the central carbon atom is known to lie on a special position on a two-fold axis (§ 6.6), the complete structure was determined even more quickly than that of beryl (p. 92).

We can thus see that optical methods provide a quick and reliable method of tackling crystal-structure problems. They can be surprisingly accurate, but the final results must always be checked and refined by computational methods.

10.5 *Optical Fourier synthesis*

We have seen in Chapter 7 that the Fourier synthesis of a crystal struc-

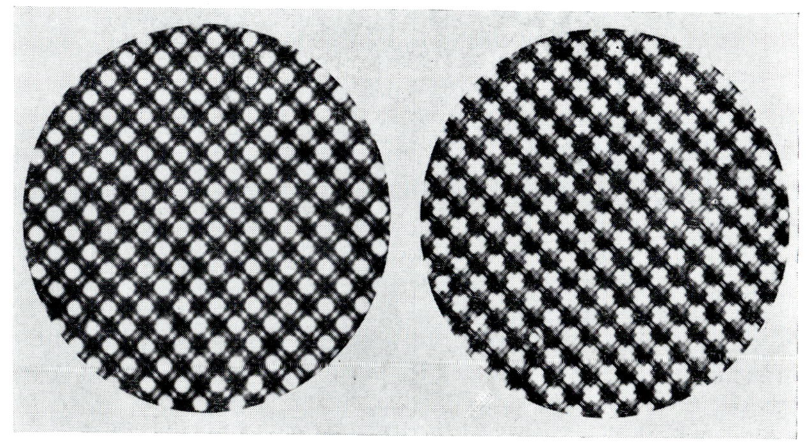

1

2

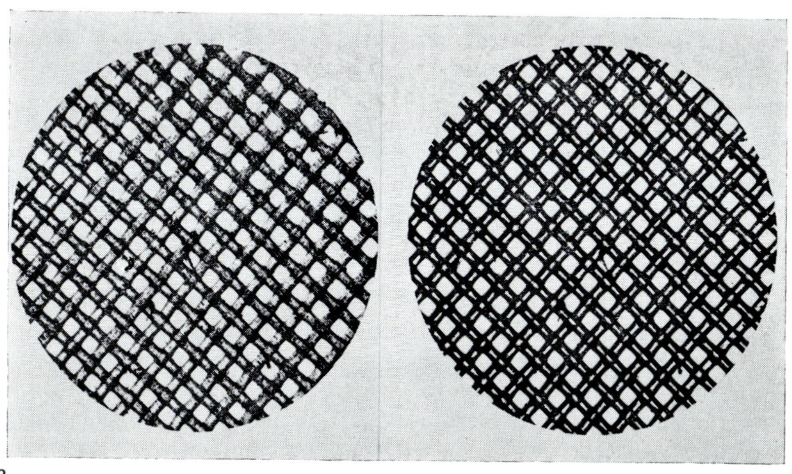

3

4

see over

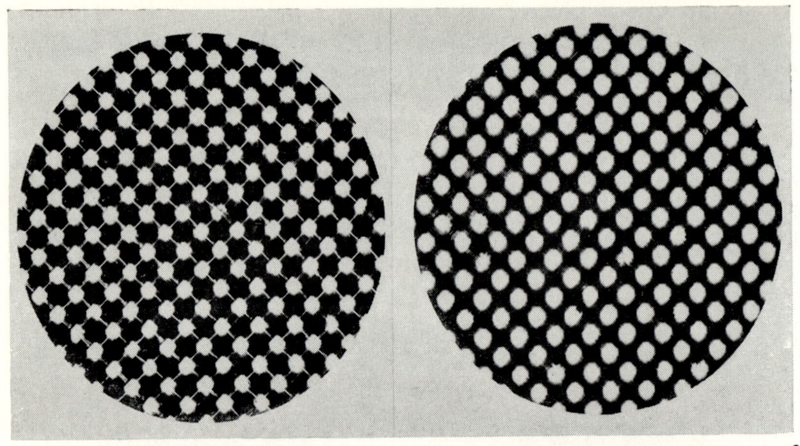

Fig. 10.11. Some out-of-focus images of a piece of gauze illuminated by a plane coherent wave.

ture is its image expressed mathematically. It is therefore tempting to ask whether the process of Fourier synthesis can be carried out experimentally by the same methods that we use to produce an optical image.

The basic obstacle is, of course, the phase problem (p. 99). To illustrate this we can use the optical diffractometer (p. 150) modified by the addition of an extra lens (fig. 10.10) which produces an image of the diffracting object. We can then see what happens to the image when it is out of focus—that is, when the relative phases of the diffracted beams are incorrect. If we use an irregular object, there is little doubt when the correct image is obtained, but if we use a regular one—such as a piece of gauze—all sorts of patterns are obtained, some of which might be thought to be correct if the detail of the object were not known. Examples are given in fig. 10.11. These experiments can easily be carried out with an ordinary microscope if the illumination is changed from the usual convergent beam to a plane wave.

Let us, however, forget the phase problem for the time being and see what we can do. Science often has to progress in this way: if an apparently insurmountable difficulty exists, one tries to see what one could do if it were not there, and the ideas so developed might possibly suggest ways round it.

The first man to explore this approach was, again, W. L. Bragg, in 1929. He considered a known structure—diopside, $CaMg(SiO_3)_2$—for which the phases could be calculated (p. 100). He used these to produce pictures of the projections of the structure upon the three principal planes, by projecting rather crude fringes—actually out-of-focus images of regularly spaced metal rods—on to photographic paper: he could control the spacing and the orientation directly and the intensity by the length of exposure; the different phases could be simulated by shifting the fringes by the required amounts. This was a crude process, but it worked surprisingly well, and it has been developed to a more routine process in an instrument called the 'Photosommateur', devised by von Eller in Paris, which is now in general use in some laboratories.

But this process is not very elegant, in that it does not make use directly of the principles of image formation—that is diffraction and interfer-

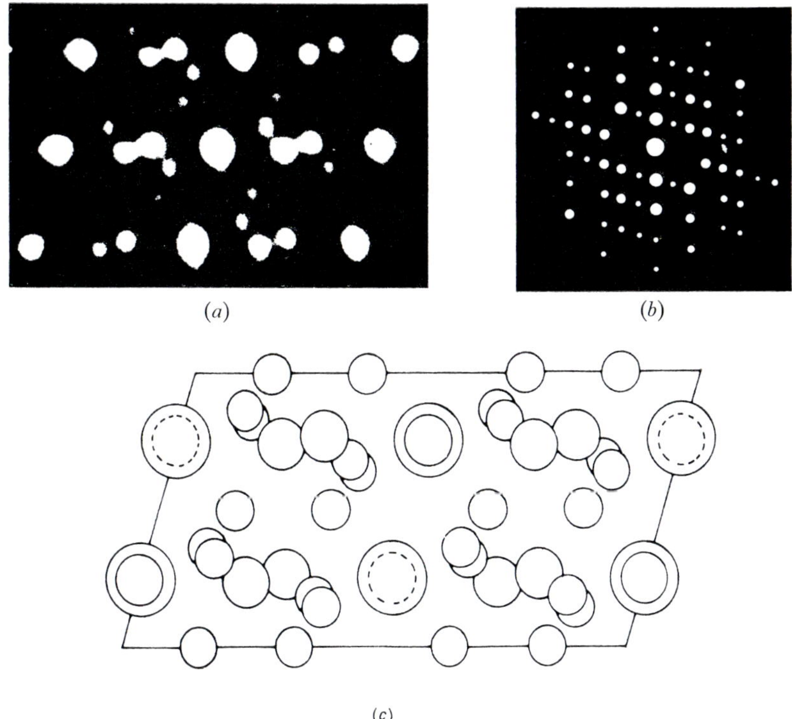

Fig. 10.12. The image (*a*) of the structure of diopside, $CaMg(SiO_3)_2$, produced from the mask shown in (*b*); (*c*) shows a diagram of the structure for comparison with (*a*).

ence (p. 102). Bragg again had another idea in 1929; could the fringes be produced by Young's double-slit interference? The different fringes could be produced by pairs of holes, whose separation should be inversely proportional to the fringe spacing, in the required orientation, and the intensity could be regulated by the sizes of the holes. Then Bragg realized that he had re-invented the reciprocal lattice!

But what about the phase problem? With this suggestion, if the reciprocal-lattice plate were illuminated by a plane wave, all the phases would be the same. So Bragg chose to try out the method with one of the projections of diopside for which the Ca and Mg atoms project on to the origin; for nearly all the reflections, these atoms, of atomic numbers 20 and 12, are sufficiently heavy to outweigh the rest of the atoms, whose atomic numbers total 76. The result, obtained with an early version of the optical diffractometer, is shown in fig. 10.12; it shows the Ca and Mg peaks clearly, and the general distribution of the Si and O atoms.

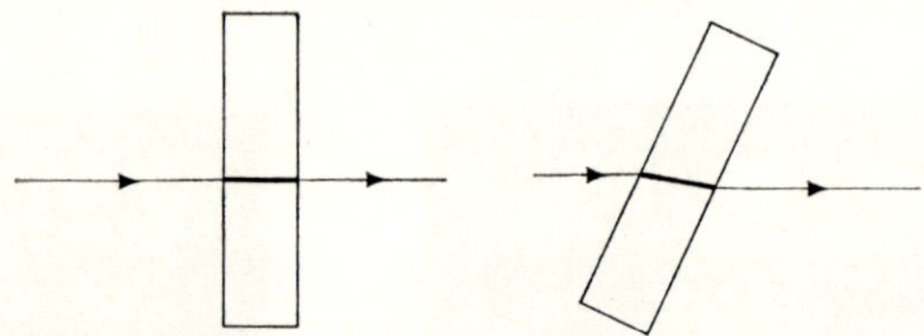

Fig. 10.13. Principle of the use of tilted mica sheets for producing phase changes. The optical path (thick line) is clearly greater for the tilted plate.

This was the key experiment that showed that the method worked. But how can we adapt it to the more general case of a projection in which all the phases are not the same? Several suggestions have been made, all depending upon the use of mica; this can be cleaved into thin sheets which are necessarily exactly uniformly thick (p. 28). Buerger in 1950 suggested using small pieces of such a sheet tilted so that their

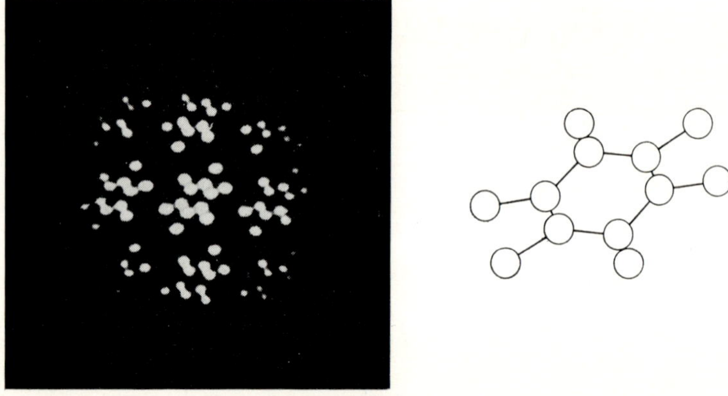

Fig. 10.14. Image of the molecule $C_6(CH_3)_6$ (p. 127) produced from the X-ray data. A drawing of the molecule is shown for comparison.

effective thickness (fig. 10.13) were such that they introduced the right path differences. More complicated devices, which will not be described here, have also been used to produce images of both centrosymmetric and non-centrosymmetric structures. The most successful is that of Hanson, Taylor and Lipson, of Manchester, who used the properties of mica in the transmission of polarized light; in 1951 they produced an image of the hexamethylbenzene molecule (fig. 10.14), which is equivalent to a direct photograph with a magnification of about a hundred million. It is not a true photograph, of course, because it is not taken in the same radiation that was diffracted by the crystal. But it is as near as we can get.

It is disappointing to have to report that these optical techniques for image formation have not really helped the subject forward. They do not contribute to solving the phase problem, and the experimentation is too 'finicky' for most crystallographers to adopt. Moreover, the results are not quantitative. But as illustrations of the physical meaning of Fourier synthesis, they are unrivalled.

CHAPTER 11
other radiations

11.1 *Introduction*

As we have seen in § 1·8, the essential property of a radiation that can be used for investigating crystal structures is that is wavelength should be about the same as the distances between atoms. X-rays with wavelengths around 1–2 Å satisfy this requirement but other radiations do so as well. In this chapter we shall consider some of the radiations that have been used and will show how they can confirm and extend the results of X-ray analysis. Two of them—electron beams and neutron beams—are now of sufficient importance to justify separate books in this series, but we also think that a book on X-ray diffraction would not be complete without a short section on the inter-relations between the different subjects.

At first, the only possible rival to X-rays were γ-rays from radioactive elements, and these were too weak to be of much use. Then in 1924 de Broglie put forward his hypothesis that moving particles should also have wave properties (p. 143), the wavelength being equal to h/mv, where h is Planck's constant and mv is the momentum. This brilliant hypothesis was found to account for some hitherto inexplicable results of Davisson and Germer and it was also tested deliberately by G. P. Thomson; they found that electron beams could be diffracted by crystals as if they were waves. Their work was later extended to protons and neutrons and the results fitted in precisely with de Broglie's hypothesis. This is now one of the bulwarks of physics and forms the basis of what we call *wave mechanics*; here, however, we shall be concerned with practical applications rather than fundamental theory.

11.2 *Electron diffraction*

The most important of the particle radiations is the electron beam; free electrons can be easily produced from a hot filament and accelerated in an electric field. An electron falling through a potential difference V acquires an energy eV, where e is its charge; this must be equal to its kinetic energy, $\frac{1}{2}mv^2$.

Thus $$\tfrac{1}{2}mv^2 = eV$$

and $$(mv)^2 = 2\,emV$$

from de Broglie's relationship, $$\lambda = h/mv,$$
$$\lambda = h/\sqrt{(2\,meV)}.$$

Putting in the values for m e and h we have that

$$\lambda \text{ in metres } = \frac{6{\cdot}63 \times 10^{-34}}{\sqrt{(2 \times 9{\cdot}11 \times 10^{-31} \times 1{\cdot}60 \times 10^{-19})} V^{-\frac{1}{2}}}$$

$$= 1{\cdot}23 \times 10^{-9}\ V^{-\frac{1}{2}}$$

$$\text{or } \lambda \text{ in Å} = 12{\cdot}3 \times V^{-\frac{1}{2}}.$$

Thus even with potential differences as low as 100 V—which is roughly what Davisson and Germer used—wavelengths of the right order of magnitude are obtained; with voltage of the order of 50 kV—which is what G. P. Thomson used—much smaller wavelengths result. Thus electrons provide a much greater range of wavelengths than X-rays do.

The scattering of electrons is quite different from that of X-rays; they are deviated by the electric fields within the atoms and, since the fields are greater in atoms of higher atomic number, the scattering factor for electrons is greater for heavier atoms, as it is for X-rays. But there is one extremely important difference; the diffraction of electrons is much stronger than that of X-rays—so strong that electron-diffraction patterns can be seen directly on fluorescent screens. X-ray diffraction patterns usually require hours of exposure; electron-diffraction patterns can be recorded in seconds.

This strength is both a disadvantage and an advantage and the former outweighs the latter. A complete theory of electron diffraction would have to take into account the diffraction of the diffracted waves, since these are almost as strong as the incident beam; in fact each order of diffraction involves an infinite converging series. Thus the theory is excessively complicated and it is not possible to work out crystal structures with electron beams as straightforwardly as it is with X-rays.

There are also some practical problems. Electrons are easily absorbed by matter and so can be transmitted only through a vacuum; we cannot direct an electron beam onto a specimen in air. For the same reason, only very thin specimens—less than about 0.1 μm—can be dealt with, and this limits the application of electron diffraction very severely. They can, however, be 'reflected' from surfaces of solid specimens. The difficulty, however, is also a source of strength. We can investigate films which are so thin that they scarcely affect X-rays. We can thus study surface effects such as oxidation and other forms of corrosion. We can even use electron beams to investigate single defects in crystals—a study that would be quite impossible with X-rays.

11.3 *Electron microscope*

However, there is a still more exciting way in which electron waves can be used. We cannot build an X-ray microscope because we cannot refract X-rays. But we *can* refract electrons, by deviating them in an electric field, and by suitably shaped electrodes we can create what are called electrostatic lenses. With these we can produce a microscope—the *electron microscope*—which has all the functions of an ordinary microscope, except that it uses electron beams instead of light. A more usual construction, because it does not need excessively high voltages, employs magnetic fields produced by electromagnets. A great deal of work, based on the theory of glass lenses and optical systems, has been put into the construction of the electron microscope, and it has now reached a high pitch of perfection.

Fig. 11.1. Image of crystal of virus, with magnification of 30 000, taken with the electron microscope (after R. W. G. Wyckoff). This shows molecules stacked in the way predicted by Huygens (fig. 2.2).

But not, unfortunately, as high as we should like. In the 1930s electron microscopes were crude instruments, with which the owners were glad to see a recognizable image. Gradually they were improved, surpassed the optical microscope in resolution, and seemed likely soon to reach atomic resolution. With wavelengths less than 0·1 Å, such resolution should have been quite possible. In fact, as limits of resolution around 10 Å were approached, practical difficulties increased; 10 Å was reached, and now 3–4 Å is claimed. But there seems no prospect of another step forward so that atomic resolution would be attained.

The electron microscope has therefore opened up a new world of observation of detail a thousand times finer than that produced by the optical microscope. It has produced images of crystals with large molecules (fig. 11.1) showing unit cells exactly in the sort of array that the classical crystallographers had deduced. It has revolutionized biology, since it shows detail inaccessible to the light microscope. But it cannot yet be used for 'seeing' atoms.

The electron microscope is not much use for detail just below the wavelength of light. This gap has now been filled by a remarkable new instrument—the *scanning electron microscope*—which can produce

Fig. 11.2. Photograph, with magnification of 10 000, of graphite crystal formation taken with scanning electron microscope. (Courtesy of I. Minkoff.)

pictures very much like those from a light microscope, but with an increase of more than ten in resolution (fig. 11.2). A beam of electrons scans a surface and the scattered electrons are made to modulate the beam of a cathode-ray tube so that a picture is produced on a television screen. This instrument has already been of immense value to metallurgists and others interested in the properties of solids.

11.4 *Neutron diffraction*

Although moving neutrons can, in principle, be used like electrons, a fundamental difficulty arises: how can we accelerate them to have the necessary momenta? Since they are uncharged, electric fields are no use. In fact, however, there is no problem; the natural thermal energy of the neutrons is sufficient to give them the velocities that we require.

In a nuclear reactor, because of the principle of equipartition of energy, the free neutrons have the same energy distribution as the rest of the atoms present. Now, the thermal energy of a particle at temperature T is $\frac{3}{2}kT$, where k is Boltzmann's constant. This must be equal to the kinetic energy, $\frac{1}{2}Mv^2$, where M is the mass of the particle. Therefore

$$(Mv)^2 = 3MkT.$$

Thus
$$\lambda = \frac{h}{\sqrt{(3MkT)}}.$$

We can work out what values of T correspond to a value of λ about the same as that of CuKα X-rays—the most widely used X-radiation. We should not be optimistic about the results of this calculation; h, M and k are extremely small quantities and we might expect to find that λ = 1·5 Å corresponds to an incredibly low or incredibly high temperature. In fact it turns out that the value is just about room temperature!

$$\lambda \text{ in metres} = \frac{6{\cdot}63\times10^{-34}}{\sqrt{(3\times1{\cdot}67\times10^{-27}\times1{\cdot}38\times10^{-23})T^{-\frac{1}{2}}}}$$

$$= 2{\cdot}51\times10^{-9}\times T^{-\frac{1}{2}}$$

For $\lambda = 1{\cdot}5\times10^{-10}$ m, $T^{\frac{1}{2}} = 25{\cdot}1/1{\cdot}5$, whence $T = 280$ K.

But our troubles are not over; $T = 280$ K gives a wide distribution of wavelengths, with a maximum at 1·5 Å, whereas we require a specific value of λ, like that of monochromatic X-rays (p. 22). In other words, we wish to select, from the neutrons travelling in all possible directions with a wide distribution of speeds, those travelling in a specific direction with a specific speed. This is easy. By passing a neutron beam through a narrow aperture (a collimator) and then reflecting it from a crystal at a chosen angle of incidence (fig. 11.3), a specific wavelength which obeys Bragg's law is selected. Such a neutron beam is said to be monochromatic. (It is a strange thought that we can associate colour with neutrons!) They can also be polarized; that is, they can all be spinning about parallel axes: but consideration of this effect would take us too far from the main subject of this chapter.

With monochromatic neutrons we can carry out the same sort of investigations as with characteristic X-rays (p. 62), although in general much bigger crystals are required. Neutron diffraction does not suffer from the theoretical difficulties that we have mentioned for electron diffraction. But there are two considerable differences. First, for nearly all atoms, the scattering of neutrons is due to the nuclei, which have dimensions of the order of 10^{-15} m (10^{-5} Å); thus these atoms be-

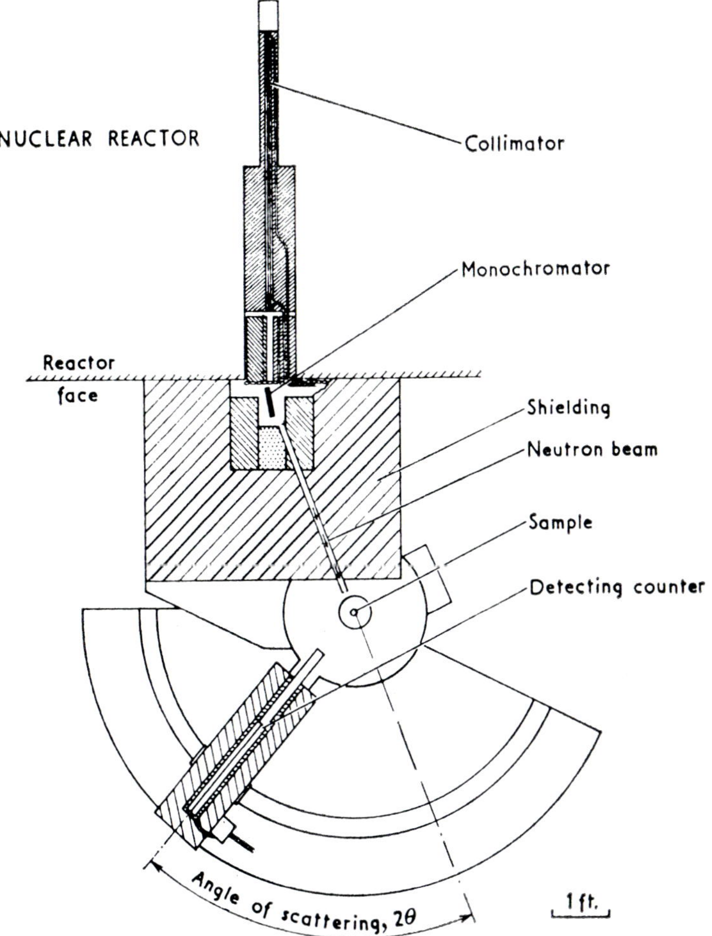

Fig. 11.3. Apparatus for obtaining neutron diffraction intensities.

have as point scatterers and their scattering factors do not decrease with angle as they do for X-rays (p. 54). Secondly, there is no direct relation between scattering factor and atomic number. Hydrogen scatters as well as potassium, for example, and deuterium is a much better scatterer still. Thus, as we shall show later, results are possible that are not obtainable with X-rays.

There are some atoms, however, for which the electrons do contribute to the scattering. These are the ferromagnetic atoms; the spinning electrons that produce the ferromagnetism can interact with the spinning neutrons. By means of this property, new information has been obtained about magnetism, and we now understand a great deal more about the subject. This research provides an excellent example of the way in which one branch of physics can help another, frequently by quite an unexpected route.

With all these advantages, the reader may ask why neutron diffraction is not replacing X-ray diffraction. The main answer is very simple; neutron beams are too expensive. Nuclear reactors are required and only the most powerful are adequate for serious research. (It was once suggested that the two halves of an atomic bomb could be placed at a suitable separation to give a useful neutron beam, but this set-up has obvious hazards!) Thus it is unlikely that neutron diffraction will ever be as extensively used as X-ray diffraction.

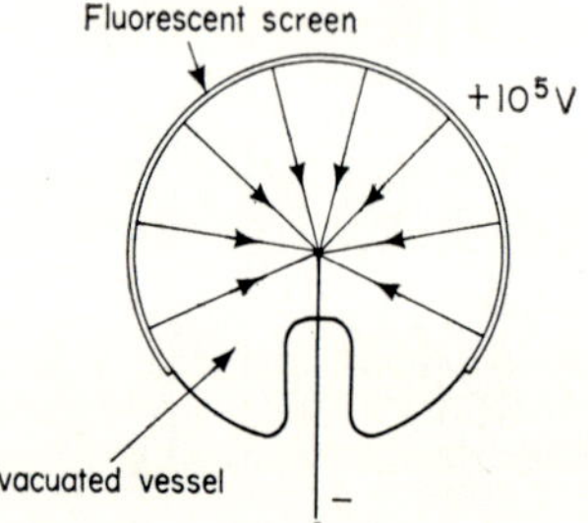

Fig. 11.4. Principle of the field-ion microscope.

11.5 *Proton beams*

It might be thought that protons would be much too damaging to be of any use for investigating crystals, since they are 1800 times as heavy as electrons. This comparison, however, is too naive. For the same momentum—which decides the wavelength—protons need move with only 1/1800 of the velocity of electrons, and therefore carry only 1/1800 of the energy.

Proton diffraction is however a relatively new subject and its uses have not been greatly explored. In fact, protons seem to behave rather in the way that the Braggs thought X-rays behaved when they believed that they were particles (p. 16); they are deviated by the atoms with which they collide and ultimately find channels that they can traverse between the atoms.

11.6 *Field-ion microscopy*

While we are discussing the use of heavy particles, it is worth while mentioning the field-ion microscope or field-emission microscope, although it does not completely fit in with the pattern of this book. It produces an image of a crystal, but not by diffraction and subsequent interference. Its importance is that it has provided the nearest approach yet to forming the images of single atoms.

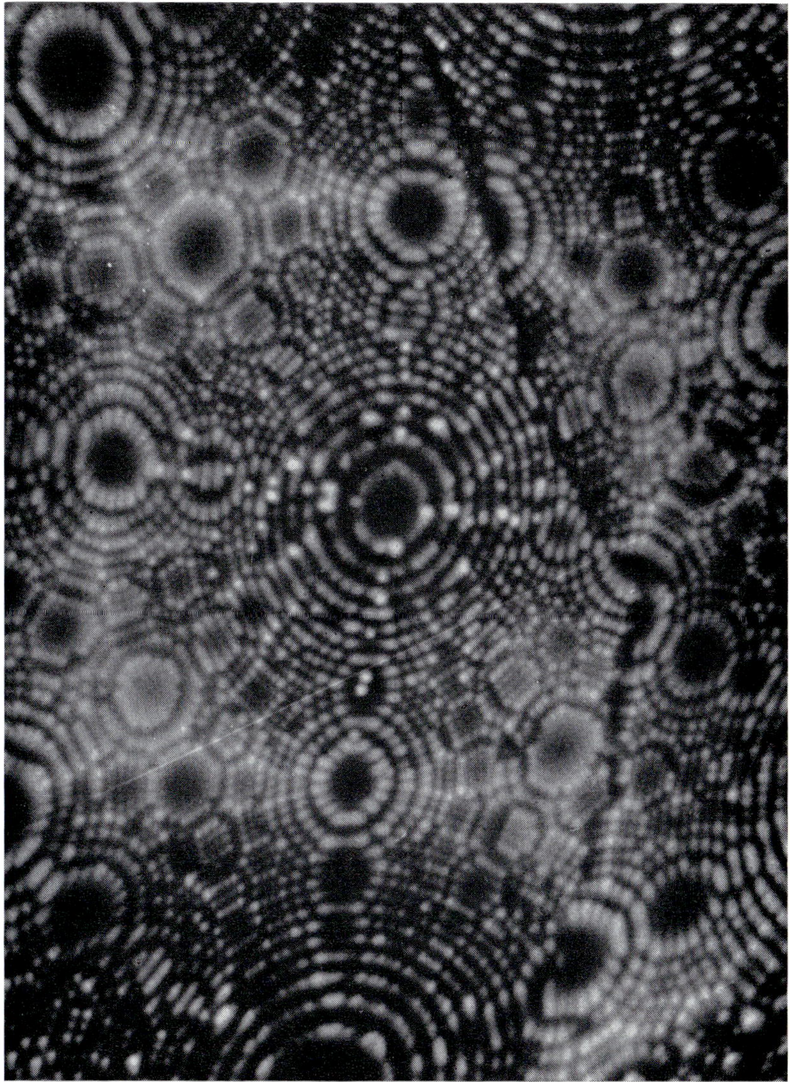

Fig. 11.5. Photograph of iridium specimen taken with field-ion microscope. (Courtesy of B. Ralph and T. F. Page)

The instrument (fig. 11.4) is extremely simple. The crystal forms a sharp point at the centre of a sphere which is coated with a fluorescent powder and is also made conducting. A high potential difference, of (a

0·50
c
Z-axis →
0·25
0
O3
½ D
½ D
O1
4(½D)
½D
½D
O2
O4,O5
-0·25
0
0·25
0·50
0·75
1·0
α sin 120°

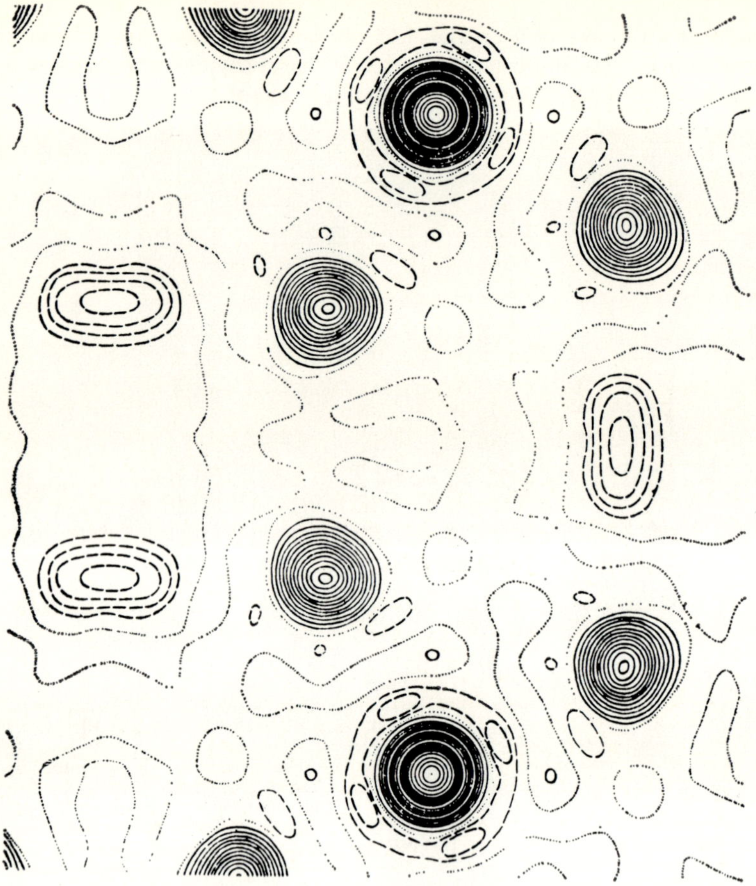

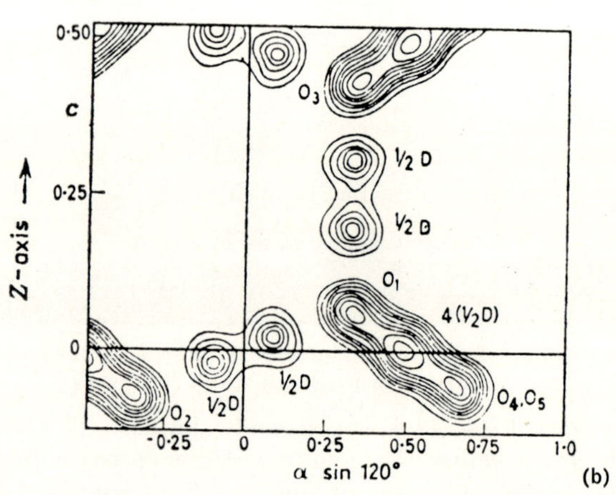

(b)

the order of 10^5 V, is maintained between the point and the sphere, and electrons are dragged from the point, travel in straight lines, and form an image on the surface of the sphere.

The action of the instrument depends upon the extremely large electric intensity near the point. Since this intensity is proportional to the number of lines of force per unit area, it increases as we approach the point (fig. 11.4); if the point has a radius of the order of 0·1 μm and the sphere a radius of 100 mm, fields of the order of 10 Vm^{-1} can be produced. Moreover, the magnification is equal to the ratio of the two radii, which, with the dimensions given, is 10^6. This compares well with the electron microscope.

The resolution is not, however, high because, amongst other things, of the large thermal motion (Brownian movement) of the atoms, which causes the electrons to have some sideways components of velocity. However, this problem can be overcome by using positive ions instead of electrons. Some helium is admitted into the sphere, and the voltage is reversed. When helium atoms collide with the crystal, they lose an electron and then travel along the lines of force to the screen. They produce pictures such as fig. 11.5 which are interpreted as showing atomic detail—not necessarily atoms themselves, but the places where atomic layers are discontinuous.

Practical applications are not yet very extensive, because of the limitations on the form of the specimen. It must usually be a very strong metal that can withstand the enormous electric fields to which it is subjected. It has been found possible to see actual faults in such crystals, but ways must be found of dealing with more general specimens before the method can make substantial contributions to the study of crystals.

11.7 *Summary*

We have now discussed the main substitutes for X-rays in examining crystalline matter, and we must now see how they all fit together. Without doubt, X-rays are predominant. They are far simpler to use and to understand than any of the other radiations; if X-ray diffraction had not been discovered first, it would have been hailed as an enormous improvement over any of its predecessors. The other techniques must therefore be considered as complementing X-rays; they should be used only for finding additional information for or dealing with problems for which X-rays are quite unsuitable.

Fig. 11.6. (*a*) Neutron-scattering density of a crystal of potassium dihydrogen phosphate. The full lines are positive contours; the broken lines are negative contours, indicating hydrogen atoms; (*b*) neutron-scattering density in a crystal of heavy ice–deuterium oxide, D_2O. The peaks labelled ½D represent positions of the deuterium atoms that are statistically only half filled.

What are these problems? First, as we have seen, X-rays are no use for examining thin surface layers; electrons are much better. Thus we must use electron diffraction or the electron microscope. X-rays cannot easily distinguish between atoms of nearly equal atomic number—for example, Cu and Zu in γ-brass (p. 90). We must therefore use neutrons to find with certainty where the two sorts of atoms are situated; the mean value of the neutron scattering factor for the two isotopes of copper, ^{63}Cu and ^{65}Cu is nearly twice as great as that for zinc.

Again, X-rays are not very good at detecting light atoms in the presence of heavy ones. In view of the general use of the heavy-atom method of crystal-structure determination, light atoms such as hydrogen are sometimes not found at all. With neutrons, however, hydrogen atoms are easily seen (fig. 11.6 *a*), and deuterium compounds are sometimes specially made to show the atoms still more clearly (fig. 11.6 *b*).

These are only some of the simplest ways in which the different radiations can help each other. For practical problems, the research worker must know how they can all be used, and must know when to call upon each to help to solve the problems with which he is faced.

CHAPTER 12
technological uses of X-ray diffraction

12.1 *General observations*

Some branches of science are entirely inward-looking. They are absorbingly interesting to the people concerned with them; but they make little impact upon other scientists, and they have no influence at all upon everyday affairs. It is, of course, inevitable and necessary that some branches of science should be like this; science could not continue if there were not some people so far ahead that only a few others can appreciate what they are doing.

Nevertheless, it is wrong that such people should be proud of the inaccessibility of their knowledge. There are several stories of great scientists who have said—one hopes in jest—'Thank God that my discoveries cannot be applied'. Scientists can justify the confidence placed in them, and the resources put at their disposal, only if their knowledge sooner or later diffuses back to the man in the street and helps him in some way to brighter and more satisfactory living.

Fortunately X-ray diffraction, right from the first years of its discovery, has been of direct help to Industry in many different ways. Industrialists have never hesitated to call in the help of X-ray diffractionists, and they, in turn, have always been glad to know that their subject, in addition to its scientific value, has had important technological applications. This chapter is concerned with describing some of these applications and discussing their importance in ordinary life.

12.2 *Identification*

By far the most important of the technological uses of X-ray diffraction is the means that it provides for identifying materials. Suppose that you are given a piece of matter and are asked to find out what it is. Most scientists would naturally turn to chemical analysis and this, of course, is the first step. But this will tell you only what elements are present, and not how they are combined. If the material is crystalline—and almost all solids are—X-ray diffraction can supplement the information given by chemical analysis, and can state definitely what compounds are present.

If the specimen is a single crystal, its unit-cell dimensions may be sufficient to identify it. All available data from crystals that have been examined have been collected together in an ordered arrangement, and any given set of cell dimensions can easily be traced. Of course, if the material has not been previously investigated the method will not work, but since over 13 000 materials have been included in the tables, and supplements are still being prepared, the chance of success is quite high.

One of the most interesting examples of single-crystal identification concerns the production of artificial diamonds. Diamond is a form of carbon, whose more usual form is graphite. It is the pre-eminent jewel and is also the hardest material known; it is therefore of considerable importance both decoratively and technologically: graphite is black and flaky, and is of importance only for certain limited applications. It would therefore be very rewarding if graphite could be turned into diamond.

In the early years of this century, claims were made that this feat had been accomplished. Because these claims seemed to smack of the alchemists' claims to have used the Philosopher's Stone to turn base metal into gold, they were not taken seriously, and X-ray methods were not brought into play. When they were thought of, in the 1940s, the original specimens could not be traced, and so we shall never know whether the claims were true or false.

The method used was to heat graphite to a high temperature in a strong small enclosure, so that very high pressures were developed. Graphite is the stable form of carbon at ordinary pressure, but it was thought possible that diamond becomes stable at high pressure, and that the existence of diamonds in the Earth is a result of the high pressures to which natural graphite has been subjected.

Although the early work was abortive, it has inspired more recent attempts. Graphite is mixed with nickel and heated to about 2000°C—well above the melting point of nickel—in a strong steel container. Pressures of the order of 10^5 atmospheres develop, and when the container is opened after cooling small particles are found in the solidified nickel. X-ray diffraction methods show that they are indeed diamond. They are not good enough for decorative purposes, but they are extremely important industrially since they enable nations that have no direct access to natural diamonds to produce their own.

X-ray powder methods can also be used for identification, and, since single crystals are relatively rare, the powder method is much more common than the single-crystal method. Even if the powder photograph of a substance cannot be interpreted, the general pattern serves as a characteristic by which the substance can be recognized. Only small samples are needed; as little as 0·1 mg is enough. For special investigations of the trans-uranic elements as little as 0·1 μg was used.

The method has often been likened to the fingerprint method of identifying people. A person's fingerprints tell us nothing about him—whether he is fat or thin, fair or dark, for example—but they can nevertheless be used to differentiate him from everyone else; in the same way, the powder photograph of a material may be too complicated to analyse, but it may differentiate that material from all others.

In order to classify powder photographs we need to present them in numerical terms. The two quantities used are, first, the spacing of the lines—($\lambda/2$) sin θ (p. 47)—and, secondly, the relative intensities. Data from photographs of about 6000 substances are arranged in order of the spacings of the strongest line in a catalogue called the A.S.T.M. (American Society for Testing Materials) Powder Data File. To identify a given powder photograph we therefore measure the spacings of the lines and pick out, by eye, the three strongest. It is easy to see whether there is any photograph in the File that has the same three strongest lines, and then to see if the rest of the pattern matches. Allowance, of course, has to be made for a certain amount of experimental error, but, if the substance is recorded in the File, there is little doubt when a match is found. Difficulties arise when the material is a mixture. If two materials are present, identification is still fairly straightforward, but with three or more constituents it becomes difficult. Nevertheless the method is so simple that it is always worth trying.

One way in which it is even superior to chemical analysis is that it tells which form of a compound occurs. For example, an important compound in paint manufacture is TiO_2, but it must be one particular form, called rutile; the other two known forms—anatase and brookite—are no use. X-ray diffraction is the most efficient way of finding out whether the paint manufacturer is buying the right compound.

The method must not, however, be given too much weight. A manufacturer of fluorescent lamps once claimed that another firm was infringing his patent by using the same chemical compound; his evidence was simply that the two gave identical X-ray powder patterns. In fact, one contained a great deal of amorphous material that gave no observable lines. The Court ruled, quite rightly, that the considerable difference in chemical composition over-rode any similarities in the diffraction patterns.

The powder method is used considerably in the steel industry. Inclusions in steel can be identified and their origin traced. Particularly is it useful in producing the correct linings for furnaces, since an incorrect ingredient can be very harmful. Here again, the ability to distinguish between the different forms of the same chemical compound —SiO_2 in the forms of quartz, cristobalite and tridymite—is extremely valuable.

The powder method of identification is of such importance in industry that some firms have installed X-ray apparatus solely for this purpose. Indeed, it is easily the most important application of X-ray diffraction and completely overshadows any other subject described in this chapter.

12.3 *Study of alloys*

There is no field in which X-ray diffraction methods of identification have had such a practical impact as in the study of metals and alloys. When two metals are melted and allowed to solidify, several possibilities may result: they may, of course, not mix at all, like oil and water, but this is very rare; one metal may 'dissolve' in the other, giving a solid solution (p. 141); the two may join together to give a new structure, called an intermetallic compound, such as Cu_5Zn_8 (p. 90); or there may be a mixture of a solid solution and an intermetallic compound or of two intermetallic compounds. Intermetallic compounds are not like chemical compounds in that they do not necessarily have fixed compositions; in other words, there may be a range of compositions over which the same structure persists. These ranges may change with temperature, and at a certain temperature the structure may disappear altogether. On the other hand, new structures may appear. The possibilities for different combinations of metals are enormous.

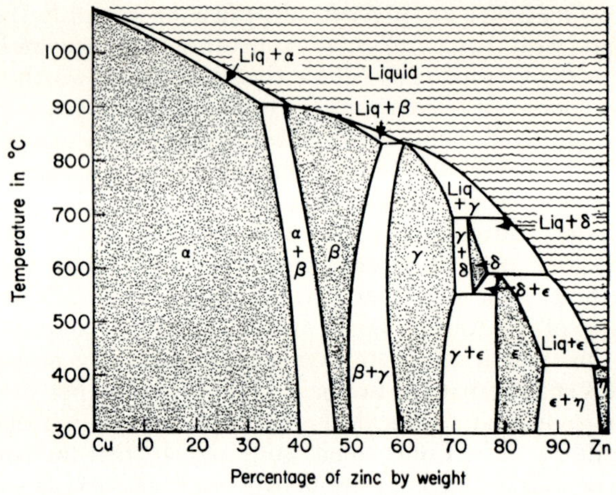

Fig. 12.1. An alloy phase diagram of moderate complexity.

Each structure is called a *phase*. A diagram showing how the phase combinations change both with composition and temperature is called a *phase diagram*. Figure 12.1 shows a typical example. The region of extent of a structure is called a *single-phase field*, and that of the coexistence of two phases is called a *two-phase field*. Three phases can coexist only at a fixed temperature and composition. The theory is similar to that of the equilibrium of water vapour, water and ice, except that, for this system, the variables are pressure and temperature; the three phases can coexist only at the triple point.

How do we build up a diagram such as fig. 12.1? The traditional way—called metallography—is to cut flat surfaces on alloys of different composition, polish them carefully until they are absolutely smooth and then etch the surface with some liquid that attacks it lightly. Examination of the surfaces under the microscope shows how many phases are present; they can be distinguished by different severity of etching, different shapes of crystal grains, or even sometimes by different colours.

The subject has developed into a fine art, and the results are often of considerable beauty—perhaps more appealing than some modern abstract art!

To find how the phase arrangements vary with temperature, the alloys may be heated to specific temperatures, quenched in cold soapy* water, and then examined in the same way. Other methods, such as cooling curves—as used ordinarily for measuring freezing points—can also be brought into play. The whole information can then be brought together to produce the sort of diagram shown in fig. 12.1.

X-ray identification methods provide an extension to metallography. Phases could be positively identified and, in addition, through the derivation of cell dimensions, could be given a measurable characteristic. In addition one could make an X-ray powder camera in which the specimen could be heated and so its structure found at high temperature. One very early piece of work—which has hardly been equalled in scope since—was the investigation by Westgren of iron; iron was thought to have four solid phases—α, β, γ and δ—at increasing temperatures. Westgren found that the α–β change was simply a change from the ferromagnetic to the paramagnetic state, the structure remaining body-centred cubic; γ (around 900°C–1400°C) is face-centred cubic; and δ (1400–1500°C) is body-centred cubic again.

One can immediately see how the application of these methods to binary alloys could effect a revolution in the subject. This is what has happened. Old problems were re-examined and often cleared up; some apparently solved problems were shown to have unexpected complications; some completely new problems were found to exist. One classical metallographer said: 'The trouble with X-ray methods is that they raise more problems than they solve.' This was meant as a criticism; the X-ray diffractionist took it as a compliment!

12.4 *Study of metal sheet and wire*

The most important property of metals is ductility. It is this which allows them to be twisted, pulled and compressed without breaking, and so enables us to manufacture the complicated shapes that we need for the articles we use in everyday life. Some metals, like copper, can be cold-worked; that is, their shapes can be changed easily at room temperature. Copper wire can be stretched by pulling it by hand, if one end is held in a vice. Other metals, like steel, need to be hot-worked; that is, they must be heated to a high temperature—usually a good red heat. This was better known to earlier generations than ours, to whom the sight of a blacksmith hammering a piece of red-hot steel was much more common than it is today.

When wire is drawn and sheet is rolled, clearly some changes must happen to the atomic arrangements. Generally, the crystals tend to be

* The lower surface tension of soapy water makes cooling much more efficient.

pulled into the same orientations (fig. 12.2), so that the material becomes more like a single crystal, and X-ray photographs show this quite clearly (fig. 12.3); instead of continuous arcs, which we should expect from a random arrangement of crystals, there are certain regions of concentration of intensity. The phenomenon is called *preferred orientation*, and has given a great deal of information about the way metals deform on the atomic scale.

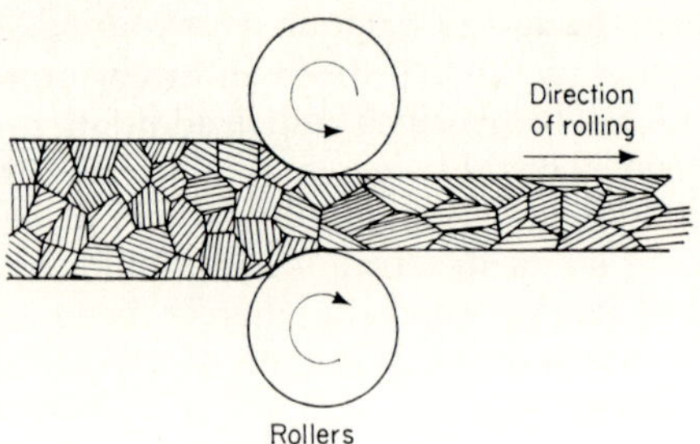

Fig. 12.2. A metal being rolled from a thicker sheet to a thinner sheet. The irregular crystals in the thicker sheet become elongated in the thinner sheet, and certain planes, represented by lines in the shading, are drawn into near parallelism.

Preferred orientation can have a great influence on the properties of the finished material, since different directions in a rolled sheet, for example, may not respond in similar ways to work performed on it. A sheet of molybdenum metal can show this effect quite clearly. In a

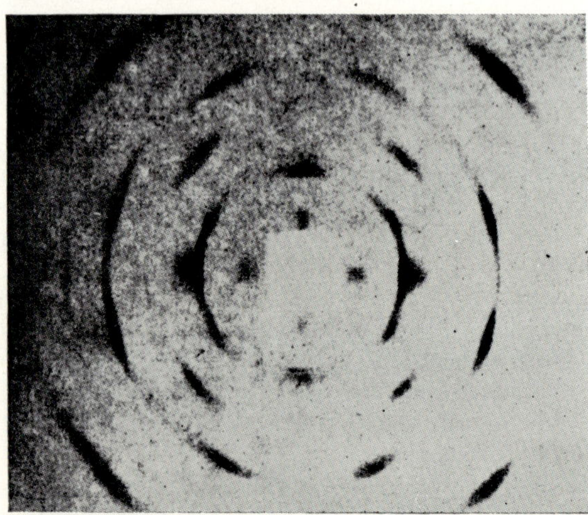

Fig. 12.3. X-ray photograph of a stationary sheet of molybdenum, showing arcs that indicate that the crystals are not randomly oriented (i.e. with preferred orientation).

certain industrial process, it was required to make some small cups by pressing them from discs cut from such a sheet. It was found, however, that the cups were not regular; they showed four 'ears' (fig. 12.4) symmetrically arranged. X-ray examination showed pronounced preferred orientation, and this had to be eliminated. A method of producing the sheet was found in which the preferred orientation was much less, X-ray methods being used to control the process at each stage.

Of course, preferred orientation is not always undesirable. In making transformers, for example, steel sheet is used in the core rather than solid metal, since this reduces the currents induced in the core (eddy currents) and consequently gives higher efficiency. In producing the sheet, preferred orientation occurs, and, if this is such that a direction of high permeability [100] is in the direction of the magnetic field, a better transformer will result. X-ray methods are used to find a heat treatment that will produce this desired effect.

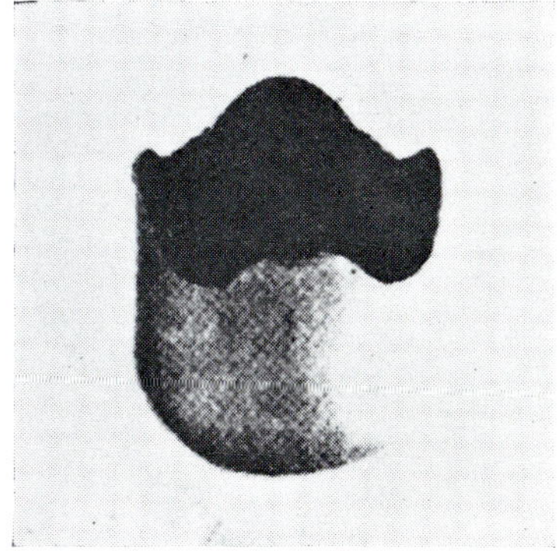

Fig. 12.4. Cups pressed from molybdenum sheet with preferred orientation, showing undesired 'ears'.

These are two of the most important examples of the use of preferred-orientation studies. There are many others that could be quoted, including the study of filaments in electric lamps. These are drawn wires and, of course, are necessarily annealed at high temperatures for long periods of time. They tend to become composed of small numbers of single crystals and breaks usually occur at the junction of two crystals. But not much use has been made of such knowledge.

12.5 *Imperfections*

Although X-ray diffraction methods have been most useful in studying perfect crystals, it has turned out that they have been able to make important technological contributions to the field of imperfectly crystalline materials. They have been of special value in dealing with metals and alloys, and it is easy to see why this should be so.

For practical purposes, pure metals are almost useless; they are

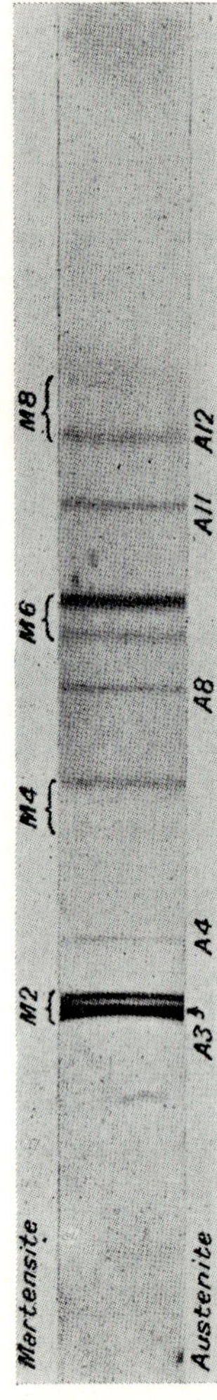

Fig. 12.5. Powder photograph of martensite, showing splitting of the lines because, for example, the spacings d_{100} and d_{001} are not equal, as they would be for a cubic compound. The lines marked M are those of martensite; those marked A are from austenite, a face-centred cubic solid solution (p. 141) of carbon in iron.

much too soft and easily deformed. Thus ornaments and medals are not made of pure platinum or gold; they are alloyed with 10–20% of another metal. The metal must be one that forms a solid solution (p. 141) with the main metal, and its purpose is to introduce deformation in the lattice, and so to prevent the easy glide (p. 139) which can take place under the action of relatively small forces.

This idea is obvious, and does not need X-ray structural investigations to support it. But there are several examples of the production of extreme hardness that were not understood until a thorough X-ray investigation had revealed the fine details of the atomic processes involved. The most important of such investigations is that of martensite—one of the phases (p. 176) that can occur in steel. Steel is made by dissolving carbon in iron and if extreme hardness is required—possibly enough to scratch glass—martensite is the phase that is sought.

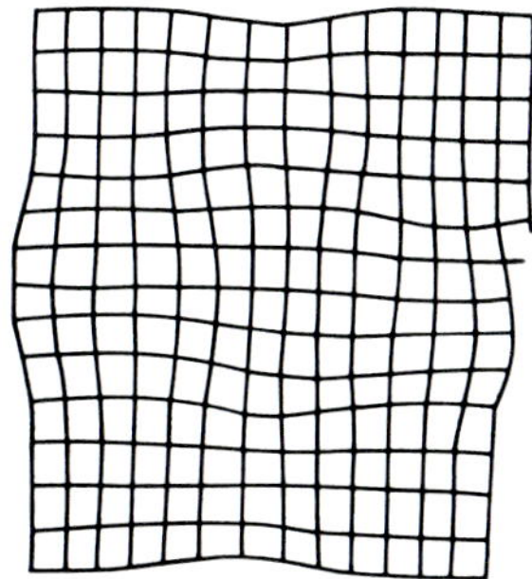

Fig. 12.6. Two-dimensional lattice built up from small regions of rectangular lattice of dimensions 5 × 6 units, in perpendicular orientations. The diagram was made by drawing several small areas of lattice, with unit cells 5 mm × 6 mm, and fitting them together, some at right angles to the others, as well as possible.

It is made by dissolving 1–1·5% of carbon in iron, cooling to 1100°C when it is solid, and then quenching abruptly in cold water.

Why should it be so hard? We know that iron at ordinary temperatures is body-centred cubic (p. 86) and experiment shows that only about 0·05% of carbon can be dissolved in it. But at 1100°C the iron is face-centred cubic and the small carbon atoms can fit more easily into the spaces between the atoms in this structure; thus over 1·5% of carbon can be dissolved. When we cool the metal quickly, the structure tries to change to body-centred cubic; the carbon atoms, however, do not have time to sort themselves out, and so become trapped in the interstices, which, as we know, are too small for them. They distort the lattice considerably, so producing great hardness.

X-ray photographs show the effect clearly. The pattern (fig. 12.5) is similar to that of body-centred cubic iron, but some of the lines are doubled. The doubling can be explained by an elongation of one of the axes with respect to the two others; the unit cell is tetragonal (p. 35). Presumably different parts of each original cubic crystal have the longer axis pointing in different directions—as shown figuratively in two dimensions in fig. 12.6—and so easy slip on the lattice planes is not possible.

Perhaps a more generally appreciated example of the importance of imperfect structures is given by aluminium. We all know how useful this metal is, and it is difficult to realize that, because of the cost of extraction, it was once regarded as a precious metal. When the electrical method of extraction made it cheap, it was still too soft to be of much use, and it was not until the early years of this century that a means was found of hardening it. The discovery—like so many of the important technological discoveries—was more or less accidental.

It was found that if an alloy with 4% copper in solution was left at room temperature, it gradually became harder. If it was left at about 200°C, the process was more rapid and the ultimate hardness greater. The alloy was hard enough to make kitchen utensils from, and the property was permanent.

The process—called *age-hardening*—was rather mysterious; it was clearly in the same class as martensite, in that the copper went into solid solution at 550°C, and then emerged in some way, but the exact details of what was happening could not be found by any ordinary method—metallographic (p. 176) or X-ray. The answer came in an unusual way—by means of single-crystal X-ray photographs.

Metallurgists had always felt that they had little use for single-crystal methods; the specimens that they were concerned with were always polycrystalline. But polycrystalline specimens are made up of single crystals, and therefore to understand them properly we must understand single crystals. This was how the problem of the age-hardening of aluminium–copper—duralumin as it was called—was solved.

Preston, in England, and Guinier, in France, both tackled the problem at the same time. They found that as the ageing proceeded, the spots on Laue photographs began to develop streaks, and that, at first, these streaks became stronger with time. They came only on one side of the spots—outward from the centre. Detailed analysis showed that the streaks lay along [100] directions in reciprocal space (p. 51), and they indicated that platelets were developing in the lattice with their normals in these directions. What could they be?

The answer that fitted in with the data was that the platelets were regions rich in copper, of composition $CuAl_2$, only a few atoms thick. The copper atoms were leaving the general solid-solution positions, and coming together in this way. If the process went on too long, the platelets would form into a definite crystal structure and then the hardness would fall. The hardness was maximum when the platelets were still fairly thin and could adhere to the basic lattice. In this way they could interfere to the maximum extent with the slip processes in the crystal, and so make it hard.

For completeness, this investigation surpasses any other that has been undertaken in this field, although it was carried out over thirty years ago. It added some new concepts to metallurgy, and, although other alloys have been found to produce similar effects, nothing essentially new has been added to the Guinier-Preston discovery.

Another field that X-ray diffraction has greatly contributed to is magnetism. To make a permanent magnet, we need an alloy of iron that is extremely hard—martensite for example. The basic principle seems to be that, to prevent the magnet slipping back after demagnetization into random orientations, some imperfections are necessary. But, until about 1930, only relatively weak magnets could be produced in this way. Moving-coil loudspeakers for radio sets had to have a separate battery for producing an electromagnet, for example, and hence they were not popular.

Then came a break-through—once again by accident. The alloy Fe_2NiAl, if it were given a certain heat-treatment, was found to be immensely superior to the best magnet previously known. Nickel is a ferromagnetic element and so seems a sensible addition, but why aluminium? X-ray photographs showed that the structure was body-centred cubic, but the lines were not at all sharp.

Bradley and Taylor, in Manchester, decided that, to solve the problem, it was no use just taking photographs of the one alloy; a wider survey was needed. They therefore undertook a complete investigation of the whole Fe–Ni–Al system and found, to their surprise, that the equilibrium structure of the alloy Fe_2NiAl was *two* body-centred cubic with different spacings. The idea that two phases in a phase diagram (p. 176) could have the *same* structure was then relatively new. The alloy is single-phase at high temperatures and as the temperature decreases it breaks up into two phases of different composition. They still try to maintain the same lattice (fig. 12.7) and the imperfections are presumably responsible for the high magnetic hardness.

Other elements, particularly cobalt, can enhance the properties, producing the well-known alloy Alnico (the iron is taken for granted!). Now many ordinary devices that need strong magnets can be made—not only moving coil loudspeakers, but bicycle dynamos, for example. No cyclist would be prepared to carry an extra battery with him to provide the magnetic field that would be needed to excite the dynamo!

Alloys such as Alnico are extremely hard and cannot be machined to suitable shapes. A softer alloy, which is still fairly good magnetically, is based upon the alloy Fe_2CuNi. This turns out to have two face-centred cubes as its equilibrium structure, but otherwise the story is similar to that of Fe_2NiAl. But the structure can be more easily controlled, and the dissociation into two phases followed closely. At first, the X-ray diffraction shows satellite lines, flanking the main lines (fig. 12.8); these are exactly similar in nature to the lines produced by an imperfectly made diffraction grating. Thus one can follow, in fair detail, the way the atoms redistribute themselves as the alloy changes from a single phase to two phases. The process is known as *spinodal decomposition*.

It is perhaps only fair to add that these technological advances would certainly have come about without the contributions of X-ray diffraction. On the other hand, it is also certain that the structural knowledge provided by X-ray diffractions has helped us to understand them and hence to control them with greater certainty than would otherwise have been possible.

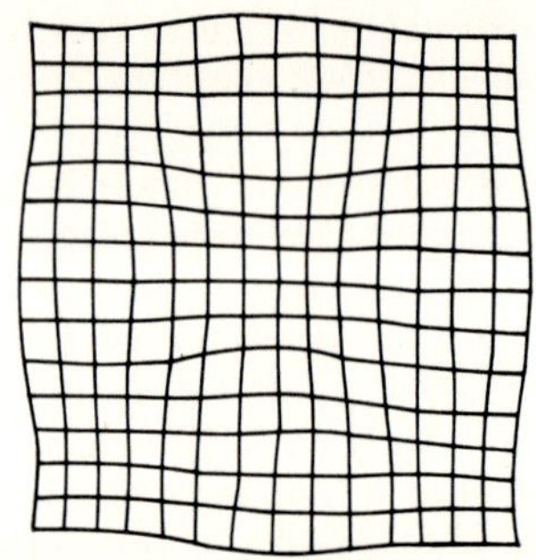

Fig. 12.7. Two-dimensional lattice built up from two square lattices, 5 × 5 and 6 × 6 units. Despite its apparent similarity to fig. 12.6, close scrutiny will show that it is essentially different.

12.6 *Crystallite sizes*

For some industrial purposes involving polycrystalline specimens, it is important to know what is called the *grain size*—the average dimensions of the crystals of which the material is composed. This may not, of course, be the same as the size of the particles themselves. X-ray diffraction gives a simple way of making this assessment.

The sample is rotated in an X-ray beam and its diffraction pattern recorded on a flat plate. If it were a single crystal, the pattern would contain relatively few spots (p. 50), but, if several crystals are present, the number of spots increases; if the number becomes very large, an ordinary powder pattern (p. 74) will be obtained. Figure 12.9 shows an example of the change from one extreme to the other.

By counting the spots in a given diffraction ring, it is possible to estimate the number of crystals in the volume irradiated if all the experimental conditions are known. Usually, however, the method is used for comparison; specimens of known crystal sizes are taken as standards, and other specimens can be compared with them. The method is very quick and certain.

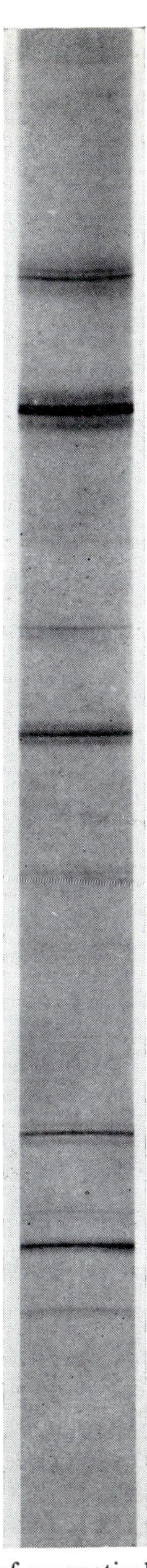

Fig. 12.8. Powder photograph of magnetically hard alloy, Fe_2NiAl, showing satellite lines produced by periodic variation of lattice spacing, as shown in fig. 12.7.

When the diffraction rings become continuous as in a normal powder photograph, the method clearly will no longer work; it can, however, be extended by keeping the specimen stationary instead of rotating it.

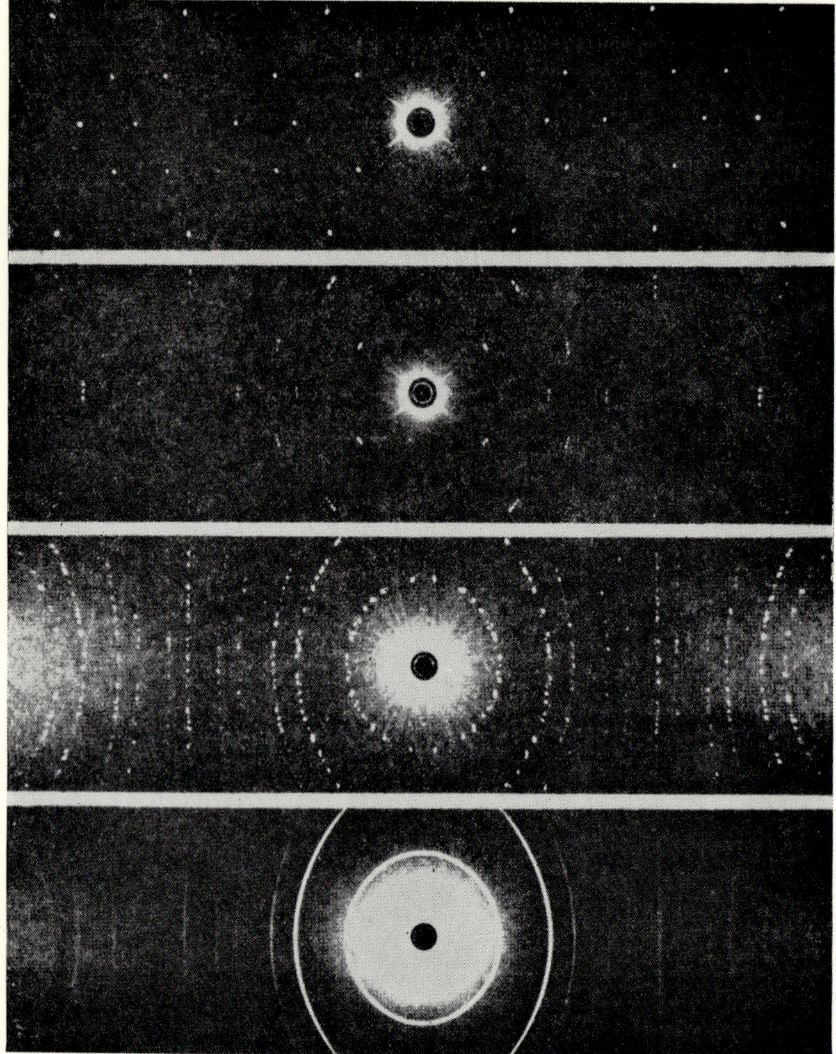

Fig. 12.9. X-ray photographs of fluorite, showing the progression from a single crystal, through a few crystals, to a large number of crystals which give a perfect powder photograph.

When the crystals become very small indeed—well below the limit at which they can be seen in a microscope—another effect becomes apparent:

the diffraction lines are no longer sharp but become blurred. The effect is exactly the same as that of reduced resolution that one obtains from a diffraction grating which consists of only a few lines. From the breadths of the lines it is possible to obtain a measure of the sizes of the crystals; and from complete measurements of the intensity as a function of Bragg angle the distribution of crystallite sizes can be obtained.

The method will work only for particles of the order of 10 nm (100 Å) or lower, and therefore it is of limited use; such small crystals are met with only rarely. But when they do occur—as in the study of colloidal suspensions—the method is probably the only reliable one.

CHAPTER 13
the future

13.1 *Implications*

As we have seen in the previous chapters, there has been immense progress since X-ray diffraction was discovered in 1912. The subject has influenced almost every branch of science in one way or another, and has revolutionized some of them. It has rewarded us with detailed information about the arrangements of atoms in solid matter, and this has allowed us to build new subjects such as solid-state physics which did not exist when X-ray crystallography began. In considering the future, therefore, it is tempting to extrapolate and to prophesy that results at least as momentous will continue to arise, and that we shall see further flowerings of science in branches which at present seem to have little relationship to X-ray diffraction as we know it.

But extrapolation is not always reliable. If a curve appears linear, do we know whether it is a straight line or the beginning of a sine curve? If we look back on the history of X-ray diffraction we can see how the subject has not always followed what seemed to be clearly indicated lines. In the following sections we shall look more closely into trends and see if there is enough evidence to indicate in what ways X-ray diffraction is going to make its chief contributions to science. We shall probably be wrong, but there is a certain amount of fun in making the attempt, and then seeing by how much we have missed the mark!

13.2 *Apparatus*

The development of X-ray apparatus provides some good examples of the dangers of extrapolation. At one period—about the 1930's—a great deal of attention was given to increasing the power of X-ray tubes; if only tubes of ten times the power could be produced, how many more problems could be tackled! The difficulty was to remove the heat produced, and X-ray tubes were made in which the target rotated so that the heated surface was continuously replaced by a cooler one. Ordinary X-ray tubes could be made with a power of 1 kW, but a gigantic apparatus was made for 50 kW. It produced some results, but basically it did not justify its existence. X-ray tubes still run at about 1 kW.

Then again, consider the sizes of cameras. The larger the radius, the better the resolution, and this is particularly important for powder cameras (p. 73). The first camera made by Hull in 1917 had a radius of 0·4 m. But it was soon discovered that smaller cameras gave shorter exposures, and diameters quickly shrank to 25 mm. Then as photographic research produced faster film, diameters increased to 50 mm, 90 mm, 190 mm (this distance was really $7\frac{1}{2}$ in!) and even 350 mm. But now diameters have settled down to about 60–190 mm; extrapolation from the 'heroic age' in the development of powder cameras would have been quite unreliable.

Nor, again, has photographic film fulfilled its early promise. At one time, there seemed to be increases of speed of the order of two every year or so, but now X-ray film does not seem to be any faster than in the 1930's.

What has happened, however, is the development of much more sensitive means of detection of X-rays—Geiger counters and such-like devices. Thus one can now produce a quantitative record of a diffraction pattern much more easily than one could from a photograph. Strangely, however, this ability has not transformed the subject; looking at a record on a piece of graph paper does not give the same impression as looking at an X-ray photograph. There is little doubt that, even if directly recording apparatus becomes still more reliable, it will not displace the old-fashioned X-ray photograph.

May there not be the possibility of incorporating developments from other branches of physics? We all know how the invention of first the maser, and then the laser, has transformed optics. The maser deals with mm waves, the laser with μm; can we apply the same techniques to X-rays which are in the nm range? To the extrapolators, the implication was obvious; the 'Xaser' was just round the corner! But the more sober scientists saw the problem more clearly; the difficulty of producing the laser action is roughly proportional to λ^{-3} and therefore we have a factor of 10^9 in going from light to X-rays. This is a not inconsiderable factor by any standards! It may be thought that if Man can reach the Moon he can do anything, but it should remembered that the distance of the Moon from the Earth is only one order of magnitude greater than the Earth's circumference.

So we can see that extrapolation with regard to apparatus would have been quite unreliable. With this fact in mind, let us see what might result if we try our abilities in other directions.

13.3 *Structure determination*

We have seen in Chapters 6 and 8 how immense success has been attained in structure determination. New methods have come into operation over the years, enabling problems of almost incredible complication to be tackled successfully. Direct methods (p. 118), coupled to improved computers, also seem to hold out hope of considerable development, and we may find that important structural information about compounds without a heavy atom (p. 108) can be obtained by their means.

But some people want to take a bigger leap still, and produce a direct image of a structure by means of a process called *holography*. This has been made possible by the existence of the laser. We have seen on p. 149 that the information about the structure of an object is contained in its diffraction pattern, but the phase problem (p. 99) prevents our retrieving this information directly. If, however, the diffraction pattern is obtained in laser light, and is superimposed upon a plane wave from the same laser (fig. 13.1) an image can be obtained by looking through the resultant interference pattern, called a *hologram*. If we could produce a 'Xaser', could we also produce an image of a crystal structure?

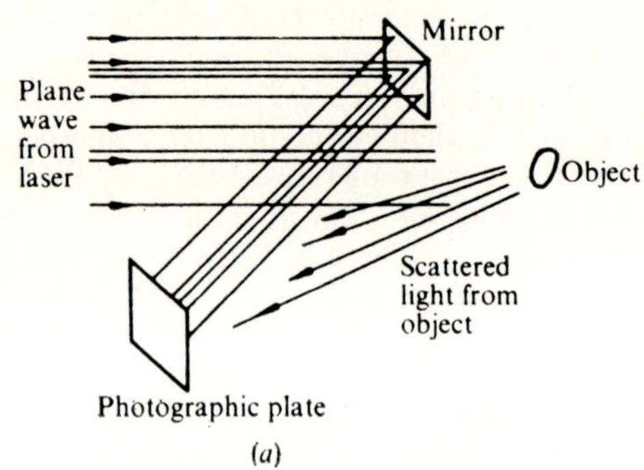

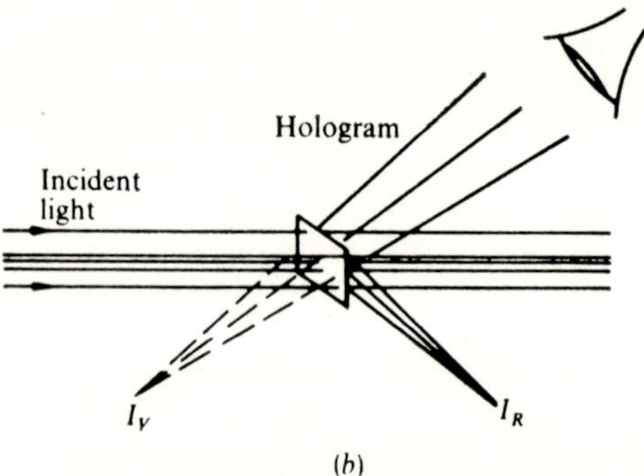

Fig. 13.1. (*a*) Set-up for producing holograms; (*b*) production of image from hologram; by looking in the direction shown one can see a virtual image I_V. A real image is produced at I_R.

The answer is 'no'. Some people doubt the wisdom of ever giving such a definite negative answer in science, but here the theory is as clear as it can be. Holography will work only if there is a continuous diffraction pattern from an object, produced all at the same time. But a crystal does not produce a continuous diffraction pattern; it produces a set of discrete orders and each one is produced with a different setting of the crystal. So, even if we could produce a Xaser, the conditions for producing an image by holography do not obtain.

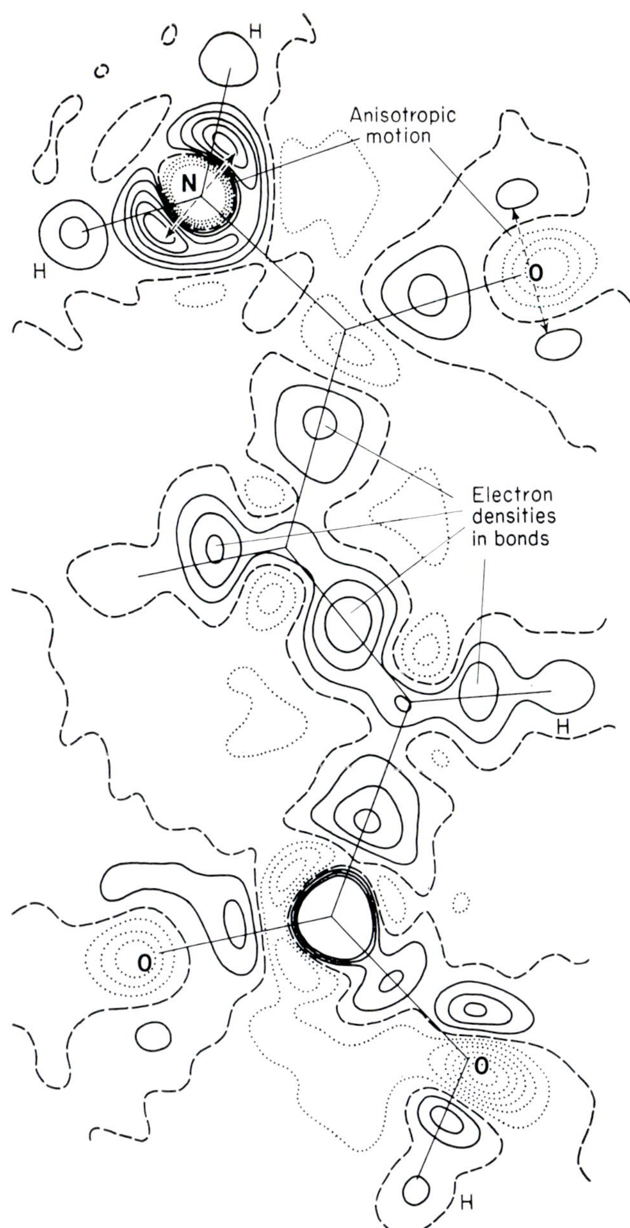

Fig. 13.2. Fine detail in electron-density map of fumaramic acid (H_2NCO-$CHCOOH$). This shows the following detail: (i) Some of the electron density in the N atom is displaced to the sides—anisotropic motion; (ii) one of the oxygen atoms shows a similar effect; the other shows it slightly; (iii) small peaks show the H atoms; (iv) electron densities are shown in the bonds, higher than in the H atoms. (Courtesy of F. Hirshfeld.)

Thus it seems unlikely that there will be any revolutionary new methods for determining crystal structures, but, as we have seen, the methods available are now so successful that the lack of anything completely new need not greatly concern us.

13.4 *Crystal structures*

In the light of the immense range of complexity of structural knowledge gained by the methods described in Chapter 8, it is tempting to think that, in the future, this range will be still further extended. This, however, is unlikely, for the reasons stated on p. 189. What is more likely is that the tremendous complications of the proteins will serve as a kind of landmark which will indicate a limit to which other researchers can aspire. Within this limit lies an extremely large number of chemical compounds.

Structural work will, however, have to be chosen with discrimination. It will soon no longer be permissible to investigate a crystal structure merely on the grounds that the result will be a contribution to knowledge; some more specific aim will be necessary—that the result will answer a chemical question or will test a new theory, for example. The days of structure determination merely for supplementing the already vast literature on bond lengths and angles are drawing to a close.

On the other hand, there is more room for investigating fine details of crystal structures—electrons in bonds, anisotropic temperature effects, imperfections in crystal packing. These investigations must take crystal-structure determination for granted, and concentrate on obtaining highly accurate measurements of X-ray intensities and means of deriving results in a meaningful way. For example, fig. 13.2 shows a representation of part of a chemical molecule in a detail that was undreamt of only a few years ago. But such work requires quite different gifts from those involved in the ordinary work of structure determination, and will attract quite different minds.

13.5 *Imperfect crystals*

In parallel with the more detailed investigation of perfect crystals must go the study of materials that are only partly crystalline. For example the polymers and plastics that are used so much in ordinary life are still far from being understood: some people maintain that they are a mixture of crystalline and non-crystalline parts; others think that their structure is homogeneous (fig. 13.3). These problems are much more intimidating than those mentioned at the end of the last section, because we do not know how to tackle them. Basically we have to explore the whole of reciprocal space not just the reciprocal-lattice points, and the phase problem (p. 99) becomes overwhelming. But we have overcome overwhelming problems in the past, and X-ray diffraction methods will not have been fully successful if they turn out to be applicable only to perfectly crystalline materials.

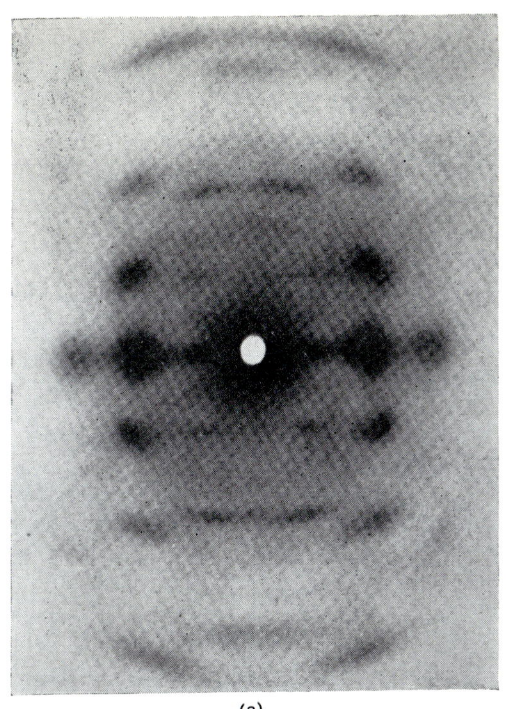

(a)

(b)

Fig. 13.3. (*a*) X-ray diffraction photograph of polymer structure; (*b*) possible structure of polymer giving rise to a pattern such as that shown in (*a*). (Courtesy of C. W. Bunn.)

13.6 *Applied crystallography*

It is difficult to see any new applications of X-ray diffraction. Without any doubt, the processes described in Chapter 12 will continue to be applied to an ever-increasing extent. There are still many industries that are not aware of the way in which they can be helped by X-ray diffraction methods; they regard X-rays and X-ray apparatus as rather exotic and do not realize how simple they are to use and how easily the results that they give can be utilized. With increasing application, however, one cannot tell what new ideas may evolve and we would therefore hope that some new application, in addition to those described in Chapter 12, would emerge.

13.7 *Penalties of success*

Thus, on the whole, it does appear that the subject of X-ray crystallography—at least as far as perfect crystals are concerned—has now reached its limit. There are many problems, but none is radically new. Progress may be likened to that of the microscope (Chapter 1). In the early days this opened up new fields and inspired a close study of lens systems and lens design. There seemed to be no limit to the performance of a microscope as techniques improved. Then came the theory of Abbe (p. 8) who showed that there *was* a limit, and when microscopes became so good that they almost achieved this limit, there was obviously no need for further development. The microscope became a standard instrument, usable by anyone who knew how to handle it, and further research in microscope design became unnecessary.

This seems to be the present state of X-ray diffraction; it has achieved the maximum of what it could reasonably be expected to do and now serves as a tool for workers in other disciplines. But just as the microscope has been revived by the invention of new devices, such as phase-contrast, so it is possible that X-ray crystallography may receive an injection of new ideas. How they will come it is impossible to say. It may be that the X-ray laser may, after all, be attained in some less direct way than by extrapolation from the laser, and this could result in an experimental solution of the phase problem. It may be that some integrated approach using diffraction by different radiations (Chapter 12) may achieve the same end. Or possibly the electron microscope (p. 164) will be so improved that atomic resolution will be possible. In forty years' time, when some of the readers of this book will be approaching the present ages of the authors, it will be possible to see how good our guesses have been. It is unlikely that we shall know!

Index

THE WYKEHAM SCIENCE SERIES

1 *Elementary Science of Metals* — J. W. MARTIN and R. A. HULL
(S.B. No. 85109 010 9) — **20s.—£1.00 net** *in U.K. only*

2 *Neutron Physics* — G. E. BACON and G. R. NOAKES
(S.B. No. 85109 020 6) — **20s.—£1.00 net** *in U.K. only*

3 *Essentials of Meteorology* — D. H. MCINTOSH, A. S. THOM and V. T. SAUNDERS
(S.B. No. 85109 040 0) — **20s.—£1.00 net** *in U.K. only*

4 *Nuclear Fusion* — H. R. HULME and A. MCB. COLLIEU
(S.B. No. 85109 050 8) — **20s.—£1.00 net** *in U.K. only*

5 *Water Waves* — N. F. BARBER and G. GHEY
(S.B. No. 85109 060 5) — **20s.—£1.00 net** *in U.K. only*

6 *Gravity and the Earth* — A. H. COOK and V. T. SAUNDERS
(S.B. No. 85109 070 2) — **20s.—£1.00 net** *in U.K. only*

7 *Relativity and High Energy Physics* — W. G. V. ROSSER and R. K. MCCULLOCH
(S.B. No. 85109 080 X) — **20s.—£1.00 net** *in U.K. only*

8 *The Method of Science* — R. HARRÉ and D. EASTWOOD
(ISBN 0 85109 090 7) — **25s.—£1.25 net** *in U.K. only*

9 *Introduction to Polymer Science* — L. R. G. TRELOAR and W. F. ARCHENHOLD
(ISBN 0 85109 100 8) — **30s.—£1.50 net** *in U.K. only*

10 *The Stars: their structure and evolution* — R. J. TAYLER and A. S. EVEREST
(ISBN 0 85109 110 5) — **30s.—£1.50 net** *in U.K. only*

11 *Superconductivity* — A. W. B. TAYLOR and G. R. NOAKES
(ISBN 0 85109 120 2) — **25s.—£1.25 net** *in U.K. only*

12 *Neutrinos* — G. M. LEWIS and G. A. WHEATLEY
(ISBN 0 85109 140 7) — **30s.—£1.50 net** *in U.K. only*

13 *Crystals and X-rays* — H. S. LIPSON and R. M. LEE
(ISBN 0 85109 150 4) — **30s.—£1.50 net** *in U.K. only*

THE WYKEHAM TECHNOLOGICAL SERIES

for universities and institutes of technology

1 *Frequency Conversion* — J. THOMSON, W. E. TURK and M. BEESLEY
(S.B. No. 85109 030 3) — **25s.—£1.25 net** *in U.K. only*

2 *The Art and Science of Electrical Measuring Instruments* — E. HANDSCOMBE
(ISBN 0 85109 130 X) — **25s.—£1.25 net** *in U.K. only*

3 *Lasers and Their Applications* — M. J. BEESLEY
(ISBN 0 85109 170 9) — **30s.—£1.50 net** *in U.K. only*

4 *Understanding Measuring Vibration* — R. H. WALLACE
(ISBN 0 85109 180 6) — **30s.—£1.50 net** *in U.K. only*